AF737590

Nuclear Power and Human Rights in Japan

Nuclear Power and Human Rights in Japan

The Fallout of Fukushima

Emrah Akyüz

LEXINGTON BOOKS
Lanham • Boulder • New York • London

Published by Lexington Books
An imprint of The Rowman & Littlefield Publishing Group, Inc.
4501 Forbes Boulevard, Suite 200, Lanham, Maryland 20706
www.rowman.com

6 Tinworth Street, London SE11 5AL, United Kingdom

Copyright © 2021 The Rowman & Littlefield Publishing Group, Inc.

All rights reserved. No part of this book may be reproduced in any form or by any electronic or mechanical means, including information storage and retrieval systems, without written permission from the publisher, except by a reviewer who may quote passages in a review.

British Library Cataloguing in Publication Information Available

Library of Congress Cataloging-in-Publication Data

Library of Congress Control Number: 2020950343

Contents

Acknowledgments

I first would like to thank my mother Kibriye Akyüz and my father Ismail Akyüz for their unfaltering support, for always believing in me and for endless encouragement and love along the way. I am deeply indebted to Akiko Tomita (Sushi) for the encouragement, motivation, continuous support, and listening ears which helped me to keep this book going throughout my fieldwork in Japan and Ukraine. Finally, I would like to express my thanks and appreciation to my supervisors, Dr. Monica Di Gregorio and Dr. James Van Alstine, for giving their time and academic support, perceptive comments, and constructive criticism about this thesis, and for their guidance throughout this challenging and thoroughly enjoyable journey.

Abbreviations and Acronyms

3Es:	Energy Security+ Economic Efficiency+ Environmental Sustainability
3Es plus S:	Energy Security+ Economic Efficiency+ Environmental Sustainability+ Safety
AEBL:	Atomic Energy Basic Law
CPHRs:	Civil and Political Human Rights
ECHR:	European Convention on Human Rights
EHRs:	Environmental Human Rights
ESHRs:	Economic and Social Human Rights
FHMS:	Fukushima Health Management Survey
FNA:	Fukushima Nuclear Accident
HR:	Human Rights
HRP:	Human Rights Policy
IAEA:	International Atomic Energy Agency
ICCPR:	International Convention on Civil and Political Rights
ICESCR:	International Covenant on Economic, Social and Cultural Rights
IEL:	International Environmental Law
IHRL:	International Human Rights Law
JAEC:	Japan Atomic Energy Commission
LDP:	Liberal Democratic Party of Japan
METI:	Ministry of Economy, Trade and Industry
mSv:	Millisievert
NEA:	Nuclear Emergency Act
NEL:	Nuclear Energy Law
NEP:	Nuclear Energy Policy
NPPs:	Nuclear Power Plants
NRA:	Nuclear Regulation Authority

PHRs: Procedural Human Rights
PR: Procedural Rights
RHPA: Radiation Hazards Prevention Act
RRA: Reactor Regulation Act
SHRs: Substantive Human Rights
TA: Thematic Analysis
TEPCO: Tokyo Electric Power Company
THUs: Temporary Housing Units
UDHR: Universal Declaration of Human Rights

Introduction

Research Background

BRIEF HISTORY OF NUCLEAR ACCIDENTS WORLDWIDE

The history of nuclear energy is relatively new, with the first grid-connected nuclear power station being the Obninsk Nuclear Power Station in Russia, which was constructed between 1951 and 1954, and produced approximately 5 MW of electric power (Semenov, 1983). Over the past sixty years, there have been a number of nuclear accidents and incidents, resulting in differing impacts and levels of danger (Christodouleas et al., 2011). The International Atomic Energy Agency (IAEA) was established as an autonomous organization in 1957 to promote the peaceful use of nuclear energy, and has developed a detailed classification of nuclear and radiological events (Forland, 1995). The International Nuclear and Radiological Event Scale (INES) categorizes nuclear accidents and incidents into the following levels: (1) nuclear and radiological events at Levels 1–3 and 4–7 are categorized as incidents and accidents, respectively; (2) events without any significance for safety are categorized Below Scale/Level 0 and are called "deviations" (i.e., anything that varies from the accepted norm or standard) (Kermisch and Labeau, 2013). The scale is designed so that each level represents an accident approximately ten times more severe than that of the previous level.

The first nuclear accident occurred on December 12, 1952, in Canada, at Chalk River, Ontario, when the NRX reactor underwent a violent power excursion that destroyed the core of the reactor, resulting in a degree of fuel melting (Thompson and Beckerley, 1973). There have subsequently been thirty-three serious incidents and accidents at nuclear power stations across the globe (Mahaffey, 2015), establishing considerable risks associated with nuclear accidents. However, it would be subjective to place all accidents and incidents into the same category, as they result in different impacts on the environment and human beings (Christodoulea et al., 2011). INES is, therefore, an effective guide for distinguishing between nuclear events, as well as a useful reference in understanding the scope and impact of the Fukushima accident in Japan.

This event scale has established two major accidents having taken place at the most dangerous level, that is, Level 7 (Kermisch and Labeau, 2013). These include: (1) the Chernobyl disaster (Soviet Union), which took place on April 26, 1986, when operators ran a test on an electrical control system of one of the reactors (Saenko et al., 2011); and (2) the Fukushima Daiichi nuclear disaster (Japan), which was primarily initiated by the tsunami triggered by the Tohoku earthquake on March 11, 2011 (Taebi and Roeser, 2015). Both these incidents have resulted in a major release of radioactive material, with widespread health and environmental implications, and have required the implementation of planned extended countermeasures (Mahaffey, 2015). Accidents at Level 7 are rare, but each poses a considerable risk to both the environment and human beings.

Accidents at Level 6 are slightly less dangerous than those at Level 7, but they also cause a significant release of radioactive material, leading to a need to implement planned countermeasures. The Kyshtym disaster at the Mayak Chemical Combine in the Soviet Union, which took place on September 29, 1957 (Cochran et al., 1993), is the only Level 6 disaster measured on the International Nuclear Event Scale. Accidents at Level 5 (unlike those at Levels 6 and 7) result in only a limited release of radioactive material, but they also require the implementation of a number of planned countermeasures, and can lead to several fatalities from radiation. Such Level 5 accidents include Three Mile Island nuclear power plant in the United States in 1979 (Walker, 2004), and Windscale in the United Kingdom in 1957 (Arnold, 1992). The final incident categorized as an accident is Level 4, which has the lowest impact on human and environment in comparison with the higher levels. Accidents at Level 4 cause a minor release of radioactive material and have local consequences. These are critical accidents causing at least one death due to radiation, for example, the Tokaimura nuclear accident in Japan in 1999 (Rothstein, 2001), the Saint-Laurent Nuclear Power Plant in France in 1969 (Hecht and Callon, 2009), and Jaslovské Bohunice in Czechoslovakia in 1977 (Hölgye et al., 1993).

Events at Levels 1, 2, and 3 are categorized as an "incident" by INES and occur with greater regularity than accidents. It is possible to rectify small incidents, but large incidents may lead to catastrophic effects. Nuclear and radiological events at Level 3 are the most serious incidents, due to the level of radiation exposure emitted being in excess of ten times the statutory annual limit for workers, resulting in nonlethal deterministic health effects.

Such incidents include the Vandellòs Nuclear Power Plant in Spain in 1989, the Paks Nuclear Power Plant in Hungary in 2003, and the THORP plant Sellafield in the UK in 2005 (Mahaffey, 2015). Events at Level 2 result in the exposure of a member of the public to a radiation dose in excess of 10 mSv (millisievert), or exposure of a worker in excess of the statutory annual limits (IAEA, 2016). Such incidents include Forsmark Nuclear Power Plant in Sweden in 2006 and Blayais Nuclear Power Plant in France in 1999 (Mahaffey, 2015). Finally, incidents at Level 1 result in the exposure of a member of the public to in excess of the statutory annual limits. These include Penly, Seine-Maritime, in France in 2012 (IAEA, 2016).

This scale can be interpreted in a number of ways. However, the classification consistently reveals that nuclear accidents and incidents have taken place from the 1950s to the 2010s, and that (despite improvements to the safety and performance of nuclear reactors over this time) the nuclear power industry is unable to prevent accidents and incidents. This has led to the conclusion that, despite the development of safety systems, there remains a considerable risk associated with nuclear accidents. One of the most important issues related to nuclear energy concerns safety (Cotton, 2015). Charles Perrow (1994), professor of Sociology at Yale University, was the first to propose a framework for characterizing complex technological systems. This was outlined in his book, *Normal Accidents: Living with High Risk Technologies*, which focused particularly on nuclear power plants (NPPs) according to their riskiness, and developed a radical approach to the safety issues relating to nuclear energy. Perrow (1994) claims that building in warnings and safeguards to ensure additional safety leads to failure, due to failures being inevitable as a result of the complexity of such systems, as confirmed by the existing nuclear accidents and incidents that have taken place despite the existence of sophisticated safety systems.

On the other hand, this scale can also be subject to criticism, as follows. Firstly, it fails to define any level beyond event Level 7, leading to an assumption that 7 is the worst level for nuclear and radiological events. However, the potential exists for nuclear accidents to exceed events at Level 7, and therefore IAEA will remain incomplete until it defines the most catastrophic of potential nuclear accident scenarios. Secondly, it is highly inconsistent to place Chernobyl and Fukushima within the same level, due to these two accidents lacking the comparability suggested by this rating. This is exemplified by a United Nation's report (published in 2005 by the Chernobyl Forum), which estimates that an accident will eventually cause approximately 4,000 deaths, the majority of these being manifested as cancer-related illnesses (ANS, 2005). However, in contrast to Chernobyl, there have been no deaths or cases of radiation sickness from the Fukushima accident (Taebi and Roeser, 2015). Similarly, contaminated areas and the evacuated areas are smaller around Fukushima, and the projected health effects in Japan are significantly lower than those following the Chernobyl accident (Steinhauser, 2014). However, Fukushima is a relatively new event, and the only Level 7 nuclear accident in the twenty-first century, and therefore will require further examination to establish any comparison with other nuclear events.

THE FUKUSHIMA DAIICHI NUCLEAR DISASTER

Following the United States and France, Japan is the third largest owner of NPPs (Steinhauser, 2014). Forty-three reactors are currently in operation, of which twenty-four are in the process of restart approvals (Kingston, 2013a). The first commercial nuclear power reactor began operating in 1966, and Japan now has more than fifty main reactors, which meet 30% of the country's electricity needs (Valentine and Sovacool, 2010). This is expected to increase to at least 40% by 2017. The total capacity of nuclear power generation in Japan has increased from 1.320 MWe in 1971 to 4.896 MWe in 2010 (Kurosaki, 2017; Taebi and Roeser, 2015). Despite being the only country in history to face the horror of nuclear bombs (i.e., Hiroshima and Nagasaki in 1945, which resulted in over 100,000 deaths (French et al., 2018), nuclear energy had become a national strategic priority in Japan until the Fukushima Daiichi nuclear disaster in 2011 (Kingston, 2013a). This was the largest nuclear accident since the Chernobyl event of 1986, and, as previously discussed, the first nuclear accident in the twenty-first century to be classified as a Level 7 event on INES, and the first disaster to include an earthquake, a tsunami, and a NPP accident (Mahaffey, 2015).

The Fukushima Daiichi Nuclear Power Station is located in the Futaba District of Fukushima Prefecture, Japan, and began operating in 1971 (Eisler, 2013). The plant is comprised of six separate boiling water reactors, originally designed by General Electric and maintained by the Tokyo Electric Power Company (TEPCO), which owns seventeen nuclear reactors (Suzuki and Kaneko, 2013). It is one of the fifteen largest nuclear power stations in the world, with a combined power capability of 5,480 MWe (Steinhauser et al., 2014).

The Fukushima Daiichi Nuclear accident was triggered by a magnitude 9.0 earthquake (which proved to be one of the largest in the history of Japan) in the Tohoku District of Japan on March 11, 2011 (Taebi and Roeser, 2015). The earthquake lasted approximately 6 min and damaged the external power supply to the nuclear reactor (Mahaffey, 2015). This is known as a "loss of offsite power" and is considered an important contributor to total risk at NPPs (Suzuki and Kaneko, 2013). However, the reactor and its backup systems were designed to handle this type of accident. Therefore, when the earthquake struck, the electricity-producing units (1, 2, and 3) shut down automatically, as designed (Taebi and Roeser, 2015). The other three units (4, 5, and 6) were not operating at the time, due to preparation for refueling, but were affected (Steinhauser et al., 2014). No serious damage was made to the reactors by the earthquake, but the power plant was vulnerable to a tsunami (Suzuki and Kaneko, 2013).

The earthquake triggered a powerful tsunami, which arrived 50 min after the initial earthquake and reached levels more than twice as high as those for which the plant was designed (NPS) (Taebi and Roeser, 2015). The tsunami damaged the external power supplies for the nuclear power station, while three units of the NPPs (i.e., Units 1, 2, and 3) sustained extensive severe core damage, which resulted in a failure of equipment, and three reactor buildings were damaged by hydrogen explosions (i.e., Units 1, 3, and 4) (Suzuki and Kaneko, 2013). Fuel in three of the reactor cores largely melted during the first three days, and radiation was released from the damaged reactors, contaminating a wide area surrounding the plant (Steinhauser et al., 2014). It is estimated that decommissioning the facility will take at least forty years (Suzuki and Kaneko, 2013).

ENVIRONMENTAL HUMAN RIGHTS (EHRS) APPROACH TO THE FUKUSHIMA NUCLEAR ACCIDENT (FNA)

Nuclear energy is a significant component of the world electricity supply in the twenty-first century. It provides about 11% of the world's electricity needs and 21% of electricity in OECD countries. Thirty countries worldwide have 437 operable civil NPPs with a further 66 under construction in 15 countries. However, nuclear energy is one of the most controversial sources of power in the world. There is a growing social and academic debate over the construction, the safety and environmental consequences of NPPs, and this concern is shared by many nuclear energy-providing countries despite the fact that international nuclear law regulates safety standards and sets international benchmarks to which states must subscribe. Recent nuclear accidents have made energy security the principal objective of nuclear energy policy (NEP) in many nations (Mahaffey, 2015; Cotton, 2015). The FNA in 2011, in particular, reignited the international debate over the safety of NPPs as it demonstrates that NPPs are particularly vulnerable to natural disasters such as tsunami and earthquake (Suzuki and Kaneko, 2013; Cotton, 2015).

The accident at the Fukushima Daiichi nuclear power plant site is still the subject of intense debate by scholars, NGOs, and international organizations as it is the most serious nuclear accident since the Chernobyl event of 1986 and the first nuclear accident in the twenty-first century to be classified as a Level 7 event on INES (Weart, 2012). Much has been written about this accident. The literature concerning the FNA, however, is mainly divided into two main categories. The first group predominantly evaluates the risks posed by the accident as a threat to environmental safety. Many analyses revolve around the notion of a "clean environment" and the specific interpretations of this offered by scholars and organizations. Scholars have generally focused on the observed and predicted environmental effects arising from the release of radioactive isotopes (Shozugawa et al., 2016). The second group of studies has addressed the human factor, investigating the impacts of the accident on public health. A growing body of empirical studies has identified an increase in the risk for specific cancers in certain subsets of the population inside the Fukushima Prefecture (Fukuda and Fukuda, 2012).

Undoubtedly, the environment and public health are the two core areas to which Fukushima will continue to pose a risk and is a likely explanation for why the literature focuses on these two aspects when discussing the accident. The environmental risks of the accident, however, are not limited to these two areas; indeed, it represents both direct and indirect risks to Fukushima residents' standards of living, such as their properties, educational conditions, places of work, liberty, access to safe water and food, and so on (Taebi and Roeser, 2015). More broadly, this interaction between the contaminated environment and the people shares three common features, which have not yet been fully explored. Firstly, many elements that have been affected by the accident, such as people's lives, health, property, and so on, are recognized as fundamental human rights at the national, regional, and international levels. For example, Articles 3, 17, and 25 of the Universal Declaration of Human Rights (UDHR), which is the first global expression of what most people consider to be the rights to which all human beings are inherently entitled, state "Everyone has the right to life, liberty and security of person"; "Everyone has the right to own property alone as well as in association with others"; and "Everyone has the right to a standard of living adequate for the health and well-being of himself and of his family, including food, clothing, housing. . .," respectively (UN, 1948). The UDHR clearly states that people have a fundamental right to life with security, health, and property, which are additionally guaranteed and protected by the Constitution of Japan. The accident, however, may pose serious risks to the enjoyment of these rights, particularly as 60,000 people are still unable to return to their homes due to the high levels of radioactive contamination that still remain—and will remain for an extended period—in the Fukushima region; there has been an increase in the rates of specific cancers, such as in the thyroid, along with an increased risk of noncommunicable diseases, including diabetes and psychological problems such as hyperactivity and emotional symptoms; the radiation levels (10 mSv) in temporary housing are much greater than the normal background levels (less than 5 mSv), which in turn poses a risk to the security of human life, as high does the radioactive contamination that might result in fatal illnesses (Taebi and Roeser, 2015). Human rights issues in relation to the environment in Fukushima are still not clear due to the lack of research in this area.

Secondly, there is a commonly held view that a safe environment is accepted as a precondition for the enjoyment of basic human rights. The term "safe environment" is a subjective judgment depending on values. However, an acceptable safety level for radioactive contamination is a radiation dose of less than 4 mSv according to international standards and countries with NPPs such as the United Kingdom and the United States. The acceptable safety level in Japan, however, is 20 mSv, which is considerably higher than the levels defined by international standards. The 1972 Stockholm Declaration was the first international document to recognize the relationship between the environment and human rights. It states that "both aspects of man's environment, the natural and the man-made, are essential to his well-being and to the enjoyment of basic human rights the right to life itself" (UN, 1972). The main rationale here comes from the fact that the environment is, according to basic need theory, the provider of fundamental human needs including air, water, and food, on which people are physically dependent. The quality of the environment may also directly influence the enjoyment of recognized rights (Cullet, 1995), which may be impossible, or at least restricted, in an unhealthy environment. On this basis, as all elements of nature including water resources, marine environment, air, and agricultural land have been heavily contaminated by the accident, the question remains as to how the contamination of the environment relates to the enjoyment of human rights in Fukushima. Thirdly, it is argued that human rights can be used in particular by victims of the accident to affect NEP, to address the issue of the contaminated environment, to force the Japanese government to achieve a safe environment, which is considered an essential tool to the realization of human rights (Anton and Shelton, 2011). These three key features link the environment and human rights under the umbrella of EHRs (Pathak, 2014), an area that is commonly ignored when discussing nuclear energy, particularly nuclear accidents. One of the main gaps in the literature is that although Fukushima seems, substantially, a human rights problem in relation to the environment, there is no fully inclusive and holistic EHRs approach to this accident. Therefore, one of the main purposes of this book is to integrate social science into the context of a legal study through the use of a qualitative case study to inform policy makers and law makers about how human rights, including the rights to life, health, and property, have been, or indeed have not been, realized or enjoyed in lieu of the absence of a safe environment resulting from the 2011 nuclear disaster.

The discussion of human rights issues in relation to the environment is not new, but the boundary between the two is still under debate. The 1972 Stockholm Declaration was the first international document to recognize a direct link between the environment and human rights. There is no commonly held definition of EHRs, but they can basically be defined as the reformulation and expansion of existing human rights and duties in the context of environmental protection (Anton and Shelton, 2011). More broadly, EHRs are a set of rights that are bestowed upon individuals, opportunities to protect the environment to ensure the preservation of associated fundamental rights, such as the right to life or health (Anderson and Boyle, 1996). These definitions encompass the links between the environment and human rights, which can be understood in two ways. Firstly, the environment is considered to be a precondition of fulfilling internationally recognized human rights, such as the rights to life and health (Pathak, 2014). The rationale here seems to stem from the fact that the environment is the most vital human requirement for existence, as it ultimately provides for all basic human needs, including water and air, which people obviously cannot live without. In the case of the Fukushima accident, the power stations damaged in this incident released radioactive materials into all elements of the environment including the air, soil, and water resources (Murakami and Oki, 2012); this means a contaminated environment that may undermine the realization or enjoyment of human rights in Japan. The question then remains: why and how are the environment and human rights related in the case of nuclear accidents?

Secondly, the legal protection of human rights (particularly procedural environmental rights) is regarded as an effective means of achieving environmental protection (Anderson and Boyle, 1996). There are three approaches being taken to mobilize human rights to address environmental issues. The first is to claim that a distinct right to the environment must be incorporated into internationally recognized rights (Kiss and Dominick, 1980). It is argued that the right to a safe environment would oblige states to secure a clean environment for their citizens (Anton and Shelton, 2012). The second key group disagrees with the first argument, contending that it is not necessary to uphold a distinct right to a safe environment because we have internationally recognized and protected human rights that can already be effectively mobilized in this regard. Prof. Alan Boyle (1996), who supports this latter view, states that the realization of legally enforceable civil and political rights, such as the right to expression, which is well protected and guaranteed under national and international law, may enable concerned groups to voice their objections to environmental protection, and make effective claims regarding environmental protection. Thirdly, moving away from a substantive definition of environmental rights, it is claimed that procedural environmental rights based on three pillars (the right to information, the right to participate, and the right to access justice) may be an effective way to protect the environment (e.g., the construction of a power plant), or policies, programs, and plans relating to the environment (May, 2013), and ensure substantive human rights (SHRs). These three vehicles, however, are not alternatives to each other; indeed, they appear to be pieces of the puzzle that aims to achieve environmental protection. It is argued that they can be effectively used by particularly concerned citizens to influence NEP and risk governance, and to address the contamination of the environment, which triggers a debate over the social acceptability of NPPs and NEP; this depends on how people perceive, tolerate, and accept the risks and benefits of NPPs.

By taking an EHRs approach to nuclear energy studies, this book has sought to answer the overarching research question: How do nuclear accidents (particularly the Fukushima event) relate to EHRs, including substantive rights and procedural rights (PRs)? By doing so, this research intends to contribute to the current knowledge in three ways. Firstly, it provides empirical evidence that contributes to the theoretical discussion as to whether there is a need for a distinct right to the environment, or whether existing human rights already require a safe environment. Secondly, it brings a certain insight to the discussion of how PRs and substantive rights are dependent on each other in terms of realizing human rights and achieving a safe environment in the matter of nuclear accidents. Finally, it investigates human rights violations in connection with the contamination of the environment caused by the FNA which, to the best of my knowledge, has not been previously considered in the literature.

RESEARCH QUESTIONS

This research has four objectives. It firstly analyzes Japan's human rights policy (HRP) and NEP in relation to EHRs. Key policy changes pre- and post-Fukushima are analyzed. It then focuses on the role of EHRs on the Fukushima nuclear energy matter. As discussed above, Anderson and Boyle (1996) point out that EHRs can be conceived in two ways: first, the environment has been, more traditionally, considered as a tool to achieve the realization of human rights. This formulates the notion that if an unsafe environment affects protected human rights by national constitutions then individuals can claim human rights violations as a result of a contaminated environment. Accordingly, the second objective aims to explore how a contaminated environment due to a nuclear accident has affected the enjoyment of human rights and whether we need a distinct right to environment or existing human rights already require safe environment. Thirdly, PRs are regarded as an effective way of addressing environmental issues by International Environmental Law

(IEL) such as the Aarhus Convention. Accordingly, the third research objective aims to explain how PRs have been used to address the contaminated environment in Fukushima and, more broadly, to affect NEP. Lastly, it aims to integrate the key empirical findings from objectives 1, 2, and 3 into Japan's HRP and argues the advantages and disadvantages of framing NEP and its possible impacts in the context of EHRs.

1.) To analyze how Japan's human rights and NEP relate to each other.
    a.) How does Japan's HRP address EHRs including civil and political rights, PRs, and the right to the environment?
    b.) How has Japan's NEP evolved since 2011 since the FNA?
    c.) How does Japan's NEP and HRP relate to each other, particularly with regards to the matter of the FNA?
2.) To investigate the relationship between the enjoyment/realization of fundamental human rights (including the right to life, health, and property) and contaminated environment raised by the nuclear accidents in Fukushima.
    a.) What are Fukushima residents' experiences of the consequences and ongoing environmental effects of the accident in relation to the right to life, health, and property?
    b.) What are the barriers to, and opportunities for, the realization and enjoyment of human rights in contaminated areas around Fukushima, particularly with regards to the right to life, health, and property?
3.) To explore the Fukushima residents' experiences of PRs in relation to the Fukushima event.
    a.) What are the experiences and attitudes of Japanese citizens toward invoking their PRs in the aftermath of the FNA?
4.) To assess the advantages and disadvantages of approaching nuclear energy through a human rights lens.
    a.) What advantages and disadvantages does an EHRs approach to nuclear energy offer, as opposed to other government regulations and other soft law related to environmental rights?

RESEARCH METHODOLOGY

This research has applied to a moderate constructivist approach, because there is a reality that the FNA happened, and has measurable, observable impacts on the environment, which can be studied. However, there is no one universal truth of EHRs that can be applied to all countries as different political, cultural, economic concepts may affect people's perspectives on EHRs. The role of the researcher is to look into the different interpretations of different groups and bring their, and our, own unique interpretations of the event in light of the related theory and literature.

A qualitative case study was chosen as the main methodology in this research because of my intention to explore the ongoing event within the real-life context of Fukushima. This research applied a single case study in Fukushima. Although Chernobyl is another Level 7 nuclear accident which had similar consequences to Fukushima, it remains a relatively old event from which to collect data, which is the reason why Fukushima is the single case study. Data was collected through semistructured interviews with more than sixty participants, as interviews enable the researcher to obtain more detailed information regarding the environmental impact of the FNA on people's human rights. Seven different groups were identified through literature review in accordance with their relevance to the research topic and the data required to achieve the research objectives. They include: (1) Fukushima residents who live in temporary housing units (THUs), (2) TEPCO managers, (3) public officials, (4) scholars who work on the FNA, (5) scholars who work on the EHRs, (6) environmental NGOs, and (7) human rights NGOs.

This research applied the thematic analysis (TA) method to analyze data because it enabled the researcher to examine narrative material from Fukushima residents' experiences of the FNA by breaking the text into relatively small units of content. Japan's human rights and energy policy documents were collected and analyzed via the document analysis method.

SUMMARY OF CHAPTERS

"Chapter 1: Conceptual Framework: Environmental Human Rights" introduces the theoretical framework applied in this research to define the research objective and questions. It further examines the theoretical discussion about human rights in connection with the environment. It represents various theoretical arguments on PRs including public participation in the decision-making process, access to information and access to justice, and SHRs including civil and political rights, right to environment, and the reinterpretation of human rights. It then argues how substantive rights and PRs are linked to each other.

"Chapter 2: Japan's Nuclear Energy and Human Rights Policies" analyzes Japan's HRP and NEP. This chapter is presented in three distinct parts. The first analyzes Japan's HRP in connection with international human rights law (IHRL) and the question of how Japan's HRP addresses human rights, the right to environment and PRs (right to public participation in the decision-making process, right to information, right to justice). Whether Japan has a HRP on human rights associated with the environment is studied. Second, it analyzes Japan's NEP pre- and post-Fukushima, and the question of what the key policy changes after the FNA are addressed. Lastly, it analyzes the extent to which, and indeed how, Japan's NEP and HRP are related to each other, particularly with regards to the matter of the FNA.

"Chapter 3: Violation of Substantive Human Rights in Fukushima" has analyzed the impacts and risks associated with the FNA on/to human rights, in particular, the right to life, health, and property through the TA of interviews with Fukushima residents. This research first discusses official reports, and previous work on the risks and impact of the accident on the rights to life, health, and property. We then analyze Fukushima residents' experiences of the consequences and ongoing environmental effects of the accident in relation to the right to life, health, and property.

"Chapter 4: Fukushima Residents' Experiences of Procedural Rights in Relation to Japan's Energy Policy" has analyzed the extent to which, and how, the Fukushima residents invoked appropriate PRs (including the right to public participation in the decision-making process, right to information, and right to justice) in lieu of the FNA and Japan's NEP through TA. Two main research questions have been studied which include: (1) What are the experiences and attitudes of Japanese citizens toward invoking their PRs in the aftermath of the FNA? (2) What are the problems and needs raised in connection with PRs in the Fukushima matter by Fukushima residents?

The main purpose of "Chapter 5: Discussion" is to interpret the major findings of this book in the light of the previous research regarding the FNA and EHRs. It consists of three parts: first, it discusses the extent to which the FNA represents a human rights problem as per its effect on the environment, which is the key research question addressed in the empirical chapters; second, it discusses the relationship between substantive rights and PRs in Fukushima, and critically evaluates the challenges and opportunities of an energy policy that recognizes public participation in the decision-making process, which is proposed as a PR by EHRs theory. Finally, some of the study's limitations are acknowledged and potential future areas of research are proposed.

"Chapter 6: The Conclusion" recapitulates this research's findings. The chapter consists of two main sections. It first attempts to summarize the major findings of this study as discussed in chapters 2, 3, and 4. The main points of each empirical and policy chapters are highlighted. Second, it considers how this research and its findings contribute to existing knowledge on the relationship between human rights and the environment, Japan's NEP and governance, and the study of nuclear accidents theoretically, methodologically, and empirically.

Chapter 1

Conceptual Framework

Environmental Human Rights

INTRODUCTION

The literature on EHRs indicates that there are two main conceptualizations of the environment within a human rights legal framework (Lewis, 2012a). Firstly, some scholars of international human rights (IHR) and environmental law note that the quality of the environment affects the enjoyment of human rights because people are dependent on nature to meet their basic needs, such as air, water, and food, for survival (Bosselmann, 2001; Anderson and Boyle, 1996; Boyd, 2011). A healthy environment is viewed as precondition of the full enjoyment of a wide range of human rights, including the right to life, health, and sanitation, because a poor-quality environment can, effectively, amount to a violation of human rights (Wolfe, 2003).

Secondly, some scholars have argued that human rights can be used effectively by concerned citizens to influence environmental policy, to address environmental issues, and to achieve environmental safety (Guillerm and Cesari, 2015; Shelton, 1991). There are three distinct human rights approaches to environmental protection: the right to the environment; reinterpretation of human rights; and PRs, including the rights to public participation in decision-making processes, access to information, and justice (Anderson and Boyle, 1996; Johnston, 1995). In these two ways, human rights and environmental protection can be construed as being connected with each other (Rajan, 2011; Lewis, 2012a).

The main purpose of this chapter is to explore the relationship between human rights and the environment through key debates in the literature. The chapter consists of two main parts. It first presents the definitions and types of human rights. Second, it introduces the definitions of human rights as pertaining to the environment, and discusses three approaches to environmental safety, including the right to environment, and the reinterpretation of human rights and PRs. It investigates the opportunities and challenges in realizing these three approaches in order to achieve a safe environment, and further examines the link between substantive rights and PRs. By doing so, the intention is to better understand the connection between human rights and the environment and, indeed, the scope of human rights approaches to the environment.

HUMAN RIGHTS THEORY

This section consists of two main parts. It first considers the definition and scope of human rights theory; it then introduces the various types of human rights, including civil and political rights, economic and social rights, and solidarity rights.

The Scope of Human Rights

The origin of human rights goes back to classical Greek and Roman thought such as that expressed through "natural right" (*lex naturalis*) and "natural law" (*jus naturale*) (Whiteside, 2012). The liberal foundations for human rights legislation can be traced back to the US Declaration of Independence of 1776 and the French declaration of the "rights of man" in 1789, which declared the "universal rights" of man (Algan, 2004; Ruppel, 2008). However, contemporary notions of human rights are relatively new. The foundation document of the global human rights regime is the Universal Human Rights Declaration of 1948 (Turner, 1993; Ishay, 2008). This represented the first universal human rights document to be unconditioned by race, sex, religion, social position, and nationality at the international level (Whelan and Donnelly, 2007; Kaletski et al., 2016).

Human rights can be viewed as moral norms which describe certain standards of human behavior (Whiteside, 2012; Ruppel, 2008; Woods, 2010). Human rights are ethical concepts which can be, and indeed are, used to measure a state's treatment of its citizens (Lorenzo, 2015; Hiskes, 2016). They reflect certain moral standards of how people should be treated, particularly by governments or other third-party groups (Schmid and Nolan, 2014). Human rights flow from the inherent dignity of the human person (Conte and Burchill, 2009; Woods, 2016). They are primarily based on individualism, meaning that a person is inherently entitled to their human rights regardless of their particular age, religion, ideology, nationality, ethnicity, gender, and so on, as a consequence of the mere fact of being human, independent of acts of law (Donnelly, 2002). In other words, individuals *unconditionally* have human rights, simply by virtue of their humanity (Turner, 1993; Ishay, 2008; Woods, 2014). Human rights aim to enhance conditions that allow people to live in freedom, justice, dignity, and economic security (Etzioni, 2010). They are, therefore, high-priority duties for states, meaning that governments have an overriding responsibility to respect, protect, and fulfill human rights (Whiteside, 2012; Woods, 2014).

As mentioned above, public authorities have three duties to enforce human rights, as human rights treaties are signed by governments (Donnelly, 2002). The first is that of respect, meaning that the government must refrain from interfering, either directly or indirectly, with the enjoyment of human rights (Sen, 2017; Conte and Burchill, 2009). The second is that of protection, which requires governments to prevent third parties from interfering in any way with the enjoyment of human rights (Risse-Kappen et al., 1999; Lorenzo, 2015). If third parties engage in practices, whether wittingly or unwittingly, which are harmful to the enjoyment of human rights, the government has an obligation to protect people's rights. The third is a promotion, which implies that a government must adopt any necessary measures and the effective use of resources to ensure the realization of human rights for everyone, especially the marginalized and the vulnerable under their jurisdiction (Lorenzo, 2015; Kaletski et al., 2016). However, the government can reduce or divert specific public expenditure which may result in the non-enjoyment of economic and social rights when an economic crisis or extreme circumstances such as war occur (Schmid and Nolan, 2014; Donnelly and Whelan, 2017).

Human rights are guaranteed by specific human right acts, constitutions, and international treaties (Jung et al., 2014; Donnelly, 2002; Woods, 2016). Only six countries have human rights acts worldwide while 195 countries guarantee human rights within their national constitutions at national levels (Donnelly and Whelan, 2017). For example, while Canada has a human rights act, Japan recognizes human rights within its constitution (Kavanagh, 2009). If a public authority breaches human rights, the individual may be able to take action under the appropriate act or constitution because they are justiciable and enforceable through the domestic court system (Renteln, 2013; Risse-Kappen et al., 1999; Woods, 2014). However, some scholars argue that human rights may not be enforceable for the following reasons: (1) lack of awareness, meaning that people may not know their rights; (2) people may find courts unreliable in terms of providing for a fair result; (3) corruption, meaning that political authority may control the country's courts; (4) lack of assertiveness, meaning that people may lack the capacity to assert themselves in court even if they have free legal assistance; and (5) the procedures to claim rights through the domestic courts are complex and highly technical, which makes it difficult for lay people to understand (Brems and

Adekoya, 2010; Collingsworth, 2002). These are just some of the obstacles to the enforcement of human rights.

Types of Human Rights

Human rights are divided into three overarching categories within the international context: socioeconomic, civil-political, and solidarity human rights, as initially proposed in 1979 by the Czech jurist Karel Vasak at the International Institute of Human Rights (Whiteside, 2012; Ruppel, 2008; Woods, 2014). While the first two allow for potential claims by individual persons against the state, the last category allows for potential claims by peoples and groups against the state. The first two are enshrined in the UDHR, as well as in several other key international and regional human rights regimes such as the International Convention on Civil and Political Rights (ICCPR) (Whelan and Donnelly, 2007).

Civil and Political Human Rights (CPHRs): First-Generation Human Rights

First-generation human rights refer to CPHRs which are embodied in the legally binding ICCPR, as signed by 148 signatory countries and ratified by Japan (House, 2014; Piovesan, 2004). ICCPR falls into two broad categories: firstly, the protection of people's freedom, life, and well-being from unwarranted infringement by the legislative and the executive powers; secondly, these rights ensure individuals can be a part of the civil and political life of the society and state without discrimination (Howard, 1983; Schwelb, 1968). CPHRs guarantee the liberty of citizens to contribute to governing the affairs of the society in which they live, either directly or through chosen representatives (Piovesan, 2004). These rights enable individuals to demand political responses to their needs and provide political opportunities for participation in decision-making processes (Cohen and Arato, 1994). In other words, CPHRs ensure that people have a voice in any decisions affecting their well-being.

CPHRs impose a negative duty on the government, meaning that citizens are not to be subjected to any action of another person or group (Hobbs and Trotter, 2013; Schwelb, 1968; Woods, 2010). These rights only require governments or social organizations to abstain from interfering with citizens' actions (Donnelly, 2002). In other words, CPHRs are respected by not preventing the individual and social organizations from pursuing a particular course of action (Conte and Burchill, 2009).

Civil human rights include the rights to freedom of opinion, speech, and expression, to freedom of religion, to freedom of the press, to freedom from torture; to a fair trial; to liberty; the right to respect privacy and family life, to freedom from discrimination, to freedom of expression, to freedom of thought, and to the freedom of movement (Conte and Burchill, 2009; Busse, 2004). Political rights are related to procedural fairness in law, including the right to seek redress or a legal remedy, to participation in civil society and politics, to asylum, to information, to boycott, to protest, to petition, to the freedom of assembly, the right to self-defense, to vote, and to public hearings (Faiz, 2016; Piovesan, 2004; Woods, 2014). These are PRs which are used to protect other rights such as economic and social human rights (ESHR).

ESHRs: Second-Generation Human Rights

Second-generation human rights refer to the ESHRs embodied in the International Covenant on Economic, Social and Cultural Rights (ICESCR) which has 145 signatory countries and has been ratified by Japan (Piovesan, 2004). The Covenant is legally binding, which obliges member states to respect, protect, and fulfill these rights (Turner, 1993; Ishay, 2008). Accordingly, ESHRs are widespread, being guaranteed by 195 countries within their constitutions (Jung et al., 2014). ESHRs aim to ensure that everybody's basic needs, such as housing, water, food, and so on, are satisfied (Donnelly, 2002). While economic rights are accrued by individuals based on their employment status, social rights protect individuals' social well-being beyond the workplace (Jung et al., 2014).

They are positive rights, which confer a duty upon the government to provide certain services (Beetham, 1995; Woods, 2010). ESHRs require governments to provide individuals with either a certain kind of good or service (Hernandez-Truyol, 1996). For example, people have the right to education, meaning that individuals have a right to free primary education. If public institutions do not build schools or do not provide any service related to education, poor communities, in particular, may not receive a proper education service, and the right to education may be compromised. For this reason, the government has a positive obligation to take appropriate steps toward the full realization of the right to education to the maximum of its available resources (Donnelly and Whelan, 2017). This is also obliged by ICESCR which requires states parties: "to take steps to the maximum of their available resources to achieve progressively the full realization of economic, social and cultural rights" (OHCHR, 1976).

Social rights include the right to highest attainable standards of mental and physical health, to property; to education, social security, and social insurance; to water; to food; to adequate standards of living including adequate food, clothing, and housing; and to the continuous improvement of living conditions (Woods, 2014; Schmid and Nolan, 2014; Jung et al., 2014).

Economic rights include the right to work under safe and healthy conditions, to rest, to leisure, to form and join trade unions, and the rights of workers to bargain collectively (Beetham, 1995; Ruppel, 2008).

Solidarity Human Rights: Third-Generation Human Rights

Third-generation human rights are collective or solidarity rights which ensure that people enjoy ESHRs and CPHRs (Algan, 2004; Downs, 1992). Some human rights theorists argue solidarity rights are essential to the enjoyment of first- and second-generation human rights (Alston, 1982; Ruppel, 2008). For example, people have a right to health, which is categorized as a second-generation human right. The right to health may not be enjoyed or realized by individuals in the absence of a clean environment as contamination may affect people's mental and physical health. Therefore, solidarity rights may help people to realize ESHRs and CPHRs.

Third-generation human rights can be both negative and positive. These rights oblige states to: (1) prevent non-state actors from interfering with the enjoyment of solidarity rights, (2) adopt the necessary measures and take steps to the maximum of their available resources to achieve the realization of solidarity rights, and (3) refrain from interfering with the enjoyment of the right to a healthy environment (Ruppel, 2008; Woods, 2009).

Third-generation human rights include the right to self-determination, to economic and social development, to a healthy environment, to peace, to the sharing of a common heritage, and to access to natural resources, to intergenerational equity and sustainability, and to humanitarian assistance (Donnelly and Whelan, 2017). However, these rights do not have any solid legal foundations at the international level. The African Charter is the only legally binding human rights document which guarantees solidarity rights including the right to a general satisfactory environment (Alston, 2013). Solidarity rights are recognized by soft law such as the 1972 Stockholm Declaration and the 1992 Rio Declaration (Donnelly, 2002).

Therefore, human rights are moral rights to which all human beings are entitled regardless of their age, gender, nationality, and so on. These rights are guaranteed by the appropriate constitutions, human rights act, and international treaties, depending on the country (or countries) in question. States have both "positive" and "negative" obligations to protect human rights, depending on the rights themselves; there are three main types of human rights including civil and political rights, economic and social rights, and solidarity rights. While states have a positive obligation to protect economic and social rights, they have a negative obligation to protect civil and political rights.

APPLYING EXISTING HUMAN RIGHTS TO ENVIRONMENTAL PROTECTION

Some scholars argue that human rights can be invoked by citizens to achieve environmental safety or address environmental issues (Anton and Shelton, 2011; Anderson and Boyle, 1996), but there is no consensus on which human rights are more effective in this regard. As argued before, environmental safety is essential to the enjoyment of human rights. Some scholars argue that such a formulation has the potential to be accepted as a human right in itself, namely, as a distinct right to the environment, which may oblige states to protect the environment as part of their positive obligation to protect human rights (Rodriguez Rivera, 2001). In contrast, other groups argue that there is no need for a distinct right to the environment because existing human rights, such as the right to life and health, already require a safe environment as a prerequisite (Anton and Shelton, 2011). Alternatively, IEL and various scholars have suggested that PRs, including the right to public participation in decision-making processes, the right to information, and the right to justice could be realized by concerned citizens in order to protect nature (Gellers and Jeffords, 2015).

This section discusses three approaches to environmental safety including the right to environment, and the reinterpretation of human rights and PRs. It discusses the opportunities and challenges in realizing these three approaches in order to achieve a safe environment.

Substantive Rights and Environmental Safety

Substantive rights involve a right to the substance of being human, including life, security, and liberty, which exists independently of all man-made laws. Substantive rights are those rights that can be protected or enforced by law. Substantive rights refer to the rights possessed by people regardless of gender, age, nationality, ideology, religion, and so on. There are two distinct SHR approaches on how to treat the environment including a distinct right to the environment and the reinterpretation of human rights (Woods, 2010). Accordingly, the following sections introduce the definitions of these rights and discuss the challenges and opportunities presented by these two approaches in realizing a safe environment.

A Distinct Right to the Environment and Environmental Safety

There is no global agreement which sets out an explicit right to a safe environment. Neither UDHR nor the ICESCR nor ICCPR reference a distinct right to the environment (Guillerm and Cesari, 2015; Lytton, 2000). However, the

1981 African Charter on Human and Peoples' Rights, the 1988 Additional Protocol to the American Human Rights Convention on Economic and Social Rights, the 2004 Revised Arab Charter on Human Rights, and the 2012 Association of Southeast Asian Nations Human Rights Declaration all recognize a distinct right to the environment (Guillerm and Cesari, 2015; Scott, 2000). Additionally, fifty-six countries guarantee the right to the environment as human rights within their constitutions (Boyd, 2011). Some of these countries include Greece, Spain, Switzerland, and Turkey, which recognize that people have the basic human right to live in a safe environment.

The right to the environment as a basic human right implies that a safe environment is one of the basic human rights to which all human beings are entitled (Thorme, 1990; Atapattu, 2002). The right to the environment indicates that people have a right to live in a safe environment because this provides their basic needs for survival, such as water to drink, air to breathe, and food to eat (Rajan, 2011; Lewis, 2012a). More specifically, the Aarhus Convention (1998) defines the right to the environment as follows: "Every person has the right to live in an environment adequate to his or her health and well-being." So, the right to the environment cannot be directly interpreted as the right to live in a pollution-free world. As stated by Hajjar Leib (2017, p. 92): "*The right to environment is not a right to an ideal environment with zero pollution or a right to pristine nature, but a right to an appropriate degree of environmental protection and conservation necessary for the enjoyment of basic human rights.*" The right to the environment, therefore, implies the right to live in an environment which is not damaging to one's health and well-being, has access to the life-support systems of the planet, generally including land, water, and air, and where the enjoyment of human rights is not otherwise violated.

There is, however, no consensus over what a safe environment actually is because people who have different values may define a safe environment in an entirely subjective manner. Broadly speaking, a safe environment refers to those conditions as providing the means and opportunity to achieve the physical and/or mental well-being of mankind (Lundgren and McMakin, 2018). A safe environment is the quality of the environment which does not threaten human health or life and the enjoyment of nature (Orji, 2012). All these definitions are somewhat vague because the perception of environmental risks and benefits is an individual and thus subjective judgment that depends on the individual's values, experiences, lifestyle, and so on, rather than being a quantitative, scientific notion (Pretty, 2007). In other words, what a safe environment actually is can vary from person to person depending on one's values (Sokolowska and Tyszka, 1995). Values can be described as a belief upon which one acts by preference. Values serve as a guiding principle for evaluating events. Values are a subjective assumption of what is right and important (Marris et al., 1996). An environmental problem can have many consequences, but which of them is deemed to be more important or risky depends on the values that people have (Pretty, 2007). Personal values vary from person to person; any number of variables can affect a person's values such as their beliefs and attitudes, past experiences, and so on (Slimak and Dietz, 2006). However, value systems can be categorized in two ways, each of which brings about different environmentally responsible behaviors, practical implications, and policies (Callicott, 1984). The first system is anthropocentrism that centers about humans and assumes the physical and mental superiority of humankind over the environment, including animals, and attaches values to different aspects of nature according to the interests of humans. The second value system is ecocentrism, which recognizes the values that nature has intrinsic value, independently of human needs or desires. Callicott (1984) argues that people who have strong ecocentric values perceive environmental risks as being greater than those who have anthropocentric values. Ultimately, the above further emphasizes the notion that environmental safety is a subjective judgment which also depends on one's value system.

Subjectivity in defining a safe environment is also commonly discussed in the literature with regards to EHRs (Proso, 2016). Many EHRs theorists agree that environmental safety is indicative of the freedom from environmental threats, which could render a person incapable of enjoying human rights such as the rights to life and health (Shelton, 1991). However, there are two different opinions about this definition. The first group argues that environmental safety can be determined through international standards, as explained above (Boyd, 2011). They have a quantitative approach in the sense that environmental safety may be regulated in an objective and quantitative manner. In contrast, the second group takes a constructivist position, in the sense that even international standards related to the radiation safety levels can be perceived differently according to one's values (Rajan, 2011; Lewis, 2012a). Anderson and Boyle (1996), for example, state that the terms adequate, safe, or healthy are ambiguous because environmental risks are a value judgment which is hard to codify in a legal framework, as discussed above. What is more certain is that a safe environment encompasses two characteristics, according to EHRs scholars. Firstly, a safe environment is only concerned with risks associated with the environment that could affect human health and well-being, rather than the nonhuman world or nature itself. That is to say, the EHRs literature does not consider biodiversity, the claims of animals, or conservation (Nickel, 1993), because HRs are fundamentally based on anthropocentric value systems (Woods, 2006). However, some argue that the human rights approach ignores (or at least underestimates) the fact that the maintenance of biological diversity and human well-being are effectively inseparable (Anton and Shelton, 2011). Mindell (2009, pp. 1563–64), for example, states that biological diversity plays an important role in human survival, including the decomposition of waste, recycling of vital nutrients, and maintenance of healthy water, air, and soil. Secondly, environmental safety is not limited to the current generation; it is also indicative of the well-being of future generations (Woods, 2016). This discussion is underpinned by sustainable development, which refers to development that meets the needs of the current generation without reducing the capacity of the environment to allow all people to live well, both now and in the future (McGoldrick, 1996).

Nevertheless, there are some quantitative standard scientific measures that can be used as general guidelines for environmental safety in terms of nuclear accident research. There is no universal standard, but a variety of epidemiology studies show that the average annual dose a person receives is less than 5 mSv/yr (Cuttler and Pollycove, 2009). For example, the largest worldwide user of NPPs, the United States, sets maximum permissible dose for individual members of the public from NPPs equivalent to 1 mSv/yr (NRC, 2018). The second largest user of NPPs, France, also sets 1 mSv/yr as the maximum admissible effective dose for the general public, though of course this does not include exposure to natural radioactivity or exposure for medical reasons. One millisievert is the limit for human activities such as electromagnetic rays. It is estimated that the French population is exposed to an average effective dose of 3.7 mSv/yr, a dose which constitutes 2.5 mSv of natural radioactivity, 1.1 mSv of medical origin, and 0.06 mSv of radiation linked to other human activities (EU, 2018). Similarly, Public Health England in the UK has calculated that, on average, people are exposed to about 2.7 mSv of radiation a year, which includes naturally occurring radiation and radiation of medical origin. According to the World Nuclear Association (2015), the average dose that people are exposed to worldwide is 2.4 mSv/yr, though this can vary quite widely depending on geographical location. The above does not mean that radiation doses below 4 mSv are completely safe; these are only regulatory limits and acceptable risk levels for radioactive contamination, according to the public authorities. The public may indeed perceive a radiation dose below 4 mSv as unsafe depending on their values; risk perception is a subjective judgment. For example, Sugimoto et al. (2014) conducted a nationwide cross-sectional random-sampled survey in March 2012, where 53% of 5,809 participants did not consider the environment to be safe due to radioactive contamination in Japan. Dose limits are only set by a regulatory body, and it is entirely possible that this may not reflect the public perception of safety. For nuclear energy research, international regulatory standards for acceptable environmental safety levels, therefore, require that the radiation that a person receives is less than 4 mSv/yr.

The maximum admissible effective dose for the general public in Japan was 1 mSv until the FNA. The Japanese government subsequently took the decision to increase the maximum yearly radiation exposure limit for the general public from 1 to 20 mSv, which is much higher than international standards or the prescribed dose limit in the United Kingdom, France, and the United States (Kingston, 2013). This increase in the annual safety limit is only the government's safety limit, but it is nevertheless entirely inconsistent with international standards; indeed, this is clearly a political decision as the choice of 20 mSv does not have any scientific explanation. It seems that the main rationale behind this change is that keeping radiation level below 20 mSv is much more achievable than 1 mSv in the contaminated area around Fukushima. This may raise the question of whether 20 mSv safe for environmental protection and human well-being is, in fact, safe. Epidemiological research shows that the risk of cancer from a single whole-body dose of 20 mSv is about 0.1% (Harada et al., 2014). However, this depends on the health, lifestyle, and so on, of the individual in question; if a person smokes, eats unhealthy food, consumes excessive quantities of alcohol, or takes irregular or no exercise, then this can increase the risk of developing cancer. That is to say, 20 mSv is the official limit, and does not reflect any scientific explanation or, indeed, the public perception of environmental safety and risk. The right to environment, therefore, refers to the right to live in an environment from which one receives a radiation dose of less than 4 mSv for nuclear energy research.

Therefore, the above shows that while there is some quantitative standard that is observable or verifiable by scientific methods at the international and national levels, the actual definition of a "safe environment" remains a subjective judgment depending on one's values. It is challenging to define what a safe environment actually ought to be because people may perceive environmental risks and benefits differently. However, this does not mean that safety levels, as determined by quantitative standards, are not safe for society. Although international and national standards might be safe enough for the people, individuals may perceive them as entirely unsafe, or at least not sufficiently safe, on the basis of their values, feelings, experiences, and beliefs. For this reason, there seems to be a need to set sociably acceptable safety and risks levels about which people do feel comfortable.

Reinterpretation of Human Rights and Environmental Safety

Some scholars argue that there is no need for a distinct right to the environment because it is already embodied within internationally established human rights such as the right to life and health (Lytton, 2000; Dreyer, 2015). As explained above, the healthy functioning of natural ecosystems is seen as essential for the enjoyment of human rights and a contaminated environment may impair and undermine human rights (Lewis, 2012b; Hiskes, 2017).

Accordingly, the environment can be seen as a further interpretative element, widening the scope of the SHRs such as the right to life, health, property, education, and so on, in a way that existing human rights require a safe environment (Cullet, 1995), as environmental degradation will clearly have a negative impact on an individual's ability to enjoy human rights (Boyd, 2012). For example, people have a right to life guaranteed by in IHRL, such as the UDHR. According to the WHO (2016), 12.6 million people die due to environmental contamination every year. This can be interpreted in such a way that a contaminated environment violates, or at the very least poses obstacles to the enjoyment of the right to life. For this reason, it can be argued that a safe environment is a prerequisite to the right to life because people may not realize their inalienable human rights in the absence of a safe environment. This interpretation is referred to as a reinterpretation of human rights by Anton and Shelton (2011), the "greening" of human rights by Boyle (2007), or the expansion of human rights by Hajjar Leib (2017), which, regardless of the terminology one employs, all imply that safe environment is already embodied in existing human rights.

This connection is explicitly recognized by various countries such as Argentina, Costa Rica, El Salvador, Greece, Kenya, Nepal, Peru, and Romania which have interpreted the right to a healthy environment as a requirement to "the inherent right to life" (Boyd, 2011). This is also recognized by the IHR regime: in General Comment No. 14, the United Nations Committee on the CESCR (Japan has ratified) elaborated on the meaning of Article 12, which states: "The State Parties to the present Covenant recognize the right of everyone to the enjoyment of the highest attainable standard of physical and mental health that right to health is intended to include . . . a healthy environment" (OHCHR, 1966). It does not, however, provide any explicit statement as to the actual rationale beyond this recognition (Lewis, 2012b). The question remains, as Boyle (2012, p. 613) notes, "Why the environment is indispensable for human rights?" or "Why should environmental protection be treated as a human rights issue?"

The literature commonly applies to basic needs theory to explain the interdependence between the two (Brei, 2013; Cullet, 1995; Anton and Shelton, 2011; Thorme, 1990). One of the most cited psychologists of the twentieth century, Abraham Maslow (1943), developed an elaborate framework to categorize innate human needs in his paper "A Theory of Human Motivation." These are displayed as a hierarchical pyramid, which is now known as "Maslow's Hierarchy of Needs." The hierarchy remains a highly popular framework across a number of different disciplines (Simons et al., 1987) and may be beneficial to the current research in identifying the interaction between the two.

The pyramid indicates that the greatest (and therefore most fundamental) level of need consists of physiological needs, which are the physical requirements for human survival, for example, air, water, and food. According to this theory, these are the most instinctive human needs and need to be met to maintain physical health. The essential premise is that should these physiological needs not be met, the human body will be unable to function correctly, with ultimately fatal results.

Maslow (1943) states that all other needs are secondary until physiological needs have been met, that is, when an individual is in need of food or water, then little else matters until he/she gains appropriate sustenance. Of the physiological needs, air is ranked as the most important element required to sustain life, without which individuals are generally unable to survive for longer than five minutes (Reader, 2006; Frey and Wilhite, 2005). The second most essential components for human survival consist of water and food, with individuals surviving for only three to four days without water (Gleick, 1996), and approximately two weeks (dependant on various factors) without food (Johnston, 2013).

However, Maslow fails to take into account the fact that the common feature of all these basic needs is the fact that air, food, and water are all provided directly or indirectly by nature. As nature is the life-giving and nurturing aspect of everything (Keles et al., 2012), it is also known as "Mother Earth" (Sacchi et al., 2013). Air, water, and soul are basic elements of "Mother Earth," meaning that human beings are physically dependent on the Earth (Blanco and Razzaque, 2009). Human life is encapsulated by the environment, which is comprised of a diversity of ecosystems, for example, climate, seas, forests, grasslands, agroecosystems, coral reefs, and freshwater systems (Keles et al., 2012). Each provides a suite of provisions that both regulate and support services that contribute to human existence, health, well-being, and livelihood (Yildiz et al., 2002). On this basis, some scholars argue that the environment seems an overarching precondition of the realization of human rights (Anton and Shelton, 2011).

The reason for the environment being vital to human rights is not only due to its provision of human needs (Thorme, 1990), but the quality of the environment may also directly influence the enjoyment of recognized human rights (Cullet, 1995; Glazebrook, 2009; Sachs, 2004). There has been vast empirical research providing evidence that environmental degradation has affected the enjoyment of human rights, particularly the right to life and health worldwide (Guillerm and Cesari, 2015). According to the associated research, climate change, for example, contributes to an increasing average global temperature and sea level and causes extreme weather events, which in turn affect the enjoyment of, in particular, the right to life, food, health, and an adequate standard of living and to property worldwide. Knox (2009, p. 482) provided evidence that climate change contributes to the rising sea levels in the Maldives, which in turn affects three basic human rights, "including the rights to life (melting ice and permafrost make travel more dangerous), property (as permafrost melts, houses collapse and residents are forced to leave their traditional homes), and health (nutrition worsens as the animals on which the Inuit depend for sustenance decline in number)." This can arguably be interpreted from the perspective that environmental contamination violates people's human rights including the rights to life, health, and property. A safe environment, therefore, seems essential to the enjoyment and realization of human rights worldwide.

To summarize, there are two main HR approaches to environmental safety. The first is the right to the environment which confers the individual the right to live in a safe environment. However, there is no consensus over what a safe environment is because environmental safety can be perceived in an entirely subjective manner, although there are a number of more definite international and national standards which regulate environmental safety in a quantitative and objective manner. This brings the challenge that the right to the environment may not be realizable because a safe environment does not, in fact, exist, or at least can be perceived entirely according to the individual. However, supporters argue that the right to the environment may oblige a country to achieve a safe environment because states have a positive duty to protect human rights. The second is that there is no need for a distinct right to the environment because existing human rights, such as the right to the life and health, already require such an environment; this perspective is referred to as the reinterpretation of human rights. Accordingly, this may oblige countries to achieve a safe environment because they are obliged to preserve those human rights which implicitly demand a safe environment. For this reason, the right to the environment does not bring any new insight into our consideration of a safe environment as current human rights already require such. However, the reinterpretation of human rights is based on an "interpretation," meaning that it does not clearly, or explicitly, state that human rights require a safe environment. Japan, for example, recognizes human rights such as the right to life but it does not recognize that these rights require a safe environment. So, while some countries or people may adopt the interpretation that existing human rights require a safe environment, others may not. However, the right to the environment clearly recognizes that people have a human right to live in a safe environment.

PRs and Environmental Safety

This section consists of three main parts. It first argues the extent to which the three PRs proposed by the Aarhus Convention and the Rio Declaration are related to human rights. Second, it introduces the scope of the three PRs including access to information, public participation in decision-making processes, and access to justice, respectively. Lastly, it argues the opportunities and challenges to PRs inherent to achieving a safe environment.

PRs as Human Rights to Ensure Environmental Safety

Broadly speaking, PRs are legal rights that govern official adjudication to secure substantive rights (Alexander, 1998). PRs are concerned with forming, administering, and enforcing substantive rights. There are many forms of PRs. However, three such rights, including the right to public participation, access to information, and access to justice, are proposed by the Aarhus Convention for effective environmental governance and which are originally rooted in human rights documents such as the UDHR. These three PRs are based on the idea that environmental decisions adequately and equitably address citizens' interests, values, assumptions, concerns, and so on (Gellers and Jeffords, 2018; Hazen, 1997). PRs are linked to democratic governance, which itself refers to the system through which people can be involved in those decisions which affect their lives (Mason, 2012; Hajjar Leib, 2017), and empower citizens, communities, and civil society groups to challenge environmental projects and influence environmental decisions and policies (Beierle, 2010). Armeni (2016) refers to these PRs as rights which are concerned with democratizing the process relating to environmental decisions, because these rights enable citizens to collectively decide how they are to be governed, or indeed how natural resources are to be governed (De Santo, 2016; Etemire, 2016).

The 1992 Rio Declaration on Environment and Development, approved by more than 150 countries, is, as an essential feature of environmental governance, the first international document categorizing procedural environmental rights (Kovar, 1993; Pallemaerts, 1992). Principle 10 of the Declaration makes a valuable contribution to the discussion of environmental rights, in stating: "*Environmental issues are best handled with the participation of all concerned citizens, at the relevant level. At the national level, each individual shall have appropriate access to information concerning the environment. . . . Effective access to judicial and administrative proceedings, including redress and remedy, shall be provided*" (UN, 1992). All these three are recognized as rights, but not necessarily as human rights. Another important international document is the Aarhus Convention, which was signed in 1998 by forty-six states and the European Union (Wates, 2005; Hartley and Wood, 2005). The Aarhus Convention draws inspiration from Principle 10 of the 1992 Rio Declaration and outlines similar principles; it also assigns individuals the rights to easily access information, participate effectively in decision making in environmental matters, and to seek justice if human rights are violated (Peeters and Nóbrega, 2014). Similarly, the 1982 World Charter for Nature, Agenda 21,

The Rotterdam Convention on the Prior Informed Consent Procedure for Certain Hazardous Chemicals and Pesticides in International Trade (art. 15), The Stockholm Convention on Persistent Organic Pollutants, and The United Nations Framework Convention on Climate Change are examples of other important environmental documents which encourage countries to adopt PRs within their national legislation (Shelton, 2006; Hajjar Leib, 2017).

IEL is, however, not the only, or indeed the first, instrument recognizing PRs (Kravchenko, 2010a). The human rights literature indicates that PRs are eventually embedded in IHRL (Boyle, 2012; Shelton, 2008). PRs find root in established human rights treaties and declarations that have since provided for generalized rights to information, participation, and justice, either explicitly or implicitly (Etemire, 2016; Boyle, 2007; Anton and Shelton, 2011). For example, Articles 19, 21, and 8 of The UDHR state: "Everyone has the right to . . . seek, receive and impart information," "Everyone has the right to take part in the government of his country, directly or through freely chosen representatives," and "Everyone has the right to an effective remedy by the competent national tribunals for acts violating the fundamental rights granted him by the constitution or by law," respectively (UN, 1948). Similarly, Articles 9, 13, and 7 of The African Charter on Human Rights state: "Every individual shall have the right to receive information"; "Every citizen shall have the right to participate freely in the government of his country, either directly or through freely chosen representatives"; and "the right to an appeal to competent national organs against acts of violating his fundamental rights as recognized and guaranteed by conventions, laws, regulations and customs in force," respectively (ACHPR, 1987). Similarly, the right to public participation in decision-making processes, access to the information and to justice is also recognized as a human right by various other international and regional human rights documents such as the International Covenant on Civil and Political Rights that Japan itself has ratified, the European Convention on Human Rights (ECHRs), and the American Convention on Human Rights (Boyle, 2012; Kravchenko, 2010b; Anton and Shelton, 2011). Some constitutional systems, such as Japan, incorporate, and make enforceable, international or regional human rights that grant PRs, making them domestically enforceable. On this basis, it can be said that first- and second-generation human rights already include PRs encompassing the right to public participation in decision-making processes, the right to information, and the right to justice, which can be used to protect SHRs in relation to the environment.

Types of PRs

This section focuses on three PRs as EHRs: the right to public participation in decision-making processes, the right to access to information, and the right to access to courts. There are three main reasons for focusing on these three particular rather than other PRs. Firstly, international and regional treaties such as the Rio Declaration and the Aarhus Convention encourage countries to adopt these PRs because they are defined as being essential to effective environmental governance. Secondly, as explained above, local, regional, and IHR courts have extended the scope of SHRs to these three particular rights. And thirdly, these three PR as EHRs are embedded in IHRLs such as UDHR and ICCPR as explained above.

Access to Information

This section has discussed six main themes in relation to access to information including: (1) its meaning and scope; (2) limitations to access to information; (3) forms of access to information; (4) timely sharing of information; (5) easily understood information; and (6) the fee for access to information, which are all addressed by the IEL documents such as the Aarhus Convention and the related literature (May, 2013; Beierle, 1998).

"Access to information" has two broad meanings. Firstly, it refers to the notion that a member of the public has the right to seek, receive, and impart scientifically proven and reliable information regarding the environment produced by a public authority or private sector institution (such as nuclear energy companies) in a lawful and timely manner, regardless of frontiers, which referred to as passive access to information (Kravchenko, 2010b). Environmental information can be disclosed in different forms including written, visual, audio, electronic, or indeed other material forms (Verestiuc and Tucaliuc, 2015). Exceptions are possible in India—an illiterate person can ask for information orally and the request will be written down by a government official (Kravchenko, 2010b). Secondly, access to information implies that public authorities should collect data and facts related to the environment and then disseminate this information to all members of the public (Mason, 2010; Kravchenko, 2010b). When public authorities and firms are obligated to develop national information systems and procedures that ensure systematic and periodic disclosure of environmental information through state-of-the-environment reports, public hearings, television, telephone, this is referred to as active access to information (Mason, 2012; Günther, 1998). This is regulated by specific law or acts that vary by country. Japan, for example, adopted its first national information disclosure law in 1999 which sets out an enforceable right of access to information in government hands.

Access to information, however, does not mean that public authorities should share *all* information related to the environment (May, 2013; Beierle, 1998). Such environmental information may include: (1) the state of the elements of the environment, such as air and atmosphere, water, soil, land, landscape, and natural sites; (2) factors, such as substances, energy, noise, radiation or waste, including radioactive waste, emissions, discharges, and other harmful impacts on the environment, which affect or will potentially affect elements of the environment; (3) measures such as policies, regulations, meeting notes, plans, programs, environmental agreements and conventions, activities, projects, and so on; and (4) the state of human health and safety and risks associated with the environment, including the contamination of the food and water chain (Verestiuc and Tucaliuc, 2015). However, information is accessible subject only to a narrow set of exceptions (Gupta, 2008). Any information related to the environment can be refused by the state (Mason, 2010). Restrictions can vary from the country to country, though the most common exemptions are, understandably, related to concerns about national security (Anton and Shelton, 2011). If the release of the requested information may adversely affect international relations, national defense, or public security, information can be refused by the public authorities under the auspices of state secrets (Anderson and Boyle, 1996; Gupta, 2010). Other exceptions include internal agency rules, commercial confidentiality, personal data protection, law enforcement records, trade secrets, commercial information pertaining to private enterprises, and so on (Kravchenko, 2010b; McKinney and Harmon, 2002). If an information request is rejected, delayed, or inadequately answered, review by a court or independent body should be available (Lemos and Agrawal, 2006; Anton and Shelton, 2011). Demand may also be legitimately refused by the public authorities if the requested information is not available (Rose-Ackerman and Halpaap, 2002).

On the other hand, disclosure of the above environmental information is not sufficient to achieve access to environmental information, because the access to information requires that members of the public can understand what is happening in the environment around them or what the environmental decision, policy, or information means with regards to the environment and their quality of life (May, 2013). The related literature suggests that public institutions should provide information in formats that are easily understandable by the public (Verestiuc and Tucaliuc, 2015). This means that information should not be unnecessarily technical, legal, or otherwise hard to understand (McKinney and Harmon, 2002). This seems particularly important in the matter of the FNA because NPPs are highly technical topics to the extent that it is unreasonable to expect lay people to understand such nuclear energy-related shared information (Akyuz, 2017). For example, the Fukushima Prefecture monitors radiation levels around the THS; this has been found to vary between 1 and 20 mSv in information which is made available on the Prefecture website (Tsubokura et al., 2012). All residents have access to this information about radiation levels (Pierpoint, 2011), but they may not know what a given radiation level means in terms of environmental safety or what the health risks associated with the measured radiation levels actually are. So, disclosure of easily understandable information is essential to the enjoyment of access to information.

The literature shows that information must be provided as soon as possible after an appropriate request is made (Kravchenko, 2010b; Verestiuc and Tucaliuc, 2015). There is, however, no consensus among scholars or indeed any universal standard as to *when* the information should be shared after the request is made (Giviashvili, 2011; May, 2013). According to the Aarhus Convention, the information requested must be provided by the appropriate public authority at the latest within one month after the request has been submitted while legislation in some countries sets a shorter deadline in this regard (Mason, 2010). However, if the information is complex and requires time to be disclosed, a date by which a response will be given must be indicated (Giviashvili, 2011; Anderson and Boyle, 1996). Additionally, the time given to respond depends on the type of information requested (Kravchenko, 2010b; Verestiuc and Tucaliuc, 2015). If an environmental disaster (e.g., nuclear accidents) that affects, or is likely to affect, people's lives occur the appropriate emergency information should be shared in the immediate aftermath of the accident (Beierle, 1998; May, 2013). This seems particularly important when a nuclear accident occurs, as even 1 h can be significant in terms of avoiding radiation-related health issues because exposure to radioactive materials for longer periods increases the risk of developing cancer-related illnesses (Cardis and Hatch, 2011). However, none of the international nuclear conventions to which Japan is party states in principle when information should be shared publicly, particularly in emergency situations.

Another important point in relation to access to the information is the associated fee (Giviashvili, 2011). As the access to the information is a right guaranteed to all members of the public, disclosure of the information must carry as little cost as possible (Mason, 2010), otherwise poorer communities, in particular, may not be able to afford it (Kravchenko, 2010b). However, it is acceptable that those who make a request for information related to the environment may be charged something (Gupta, 2010), but the amount varies from country to country (Verestiuc and Tucaliuc, 2015). According to the Aarhus Convention, reasonable fees can be charged as a rule (Beierle, 1998; May, 2013). The term "reasonable," however, can be perceived differently from person to person, as not everybody has the same or even similar economic power (Beierle, 1998; May, 2013). For example, while £10 might be reasonable to a middle-class person, it may not be reasonable for those who earn the minimum wage. In contrast to this vague definition in the Aarhus Convention, the literature provides clearer criteria in the sense that the fees required may not exceed the cost of the materials used to reproduce the information, such as photocopies of documents, in addition to the cost of sending it to those who requested it (Mason, 2010; Anderson and Boyle, 1996; Beierle, 1998).

Public Participation

This section aims to discuss three of the main features of public participation including the scope and main features of public participation and actors in the decision-making process in order, which are the most common features that the literature on PRs addresses.

The Definition and Scope of Public Participation

Broadly speaking, public participation is the practice of consulting and involving members of the public in decisions broadly speaking, including risks assessment, agenda-setting, policies, plan, and program formulation activities by the actors responsible for the projects that will potentially have an impact on the environment that people live in (Gunderson, 2018). The term public participation can be viewed as the intention or will for the redistribution of power between the government and the public when formulating policies, projects, and decisions (Konisky et al., 2001). Public participation provides the opportunity for citizens and civil society organizations to influence the decisions and programs that will likely affect them through dialogue, debate, and discussion with powerholders (Lee and Abbot, 2003) because it may allow various groups from different social backgrounds to add input to decisions and provide the opportunity for citizens to take a part in shaping of policies and plans that will affect the environment in which they live at the local, national, regional, or international levels (Verschuuren et al., 2004). More specifically, public participation may encompass three main dimensions which include the right to know (transparency), the right to comment (participation), and the right of standing (accountability) (Bishop and Davis, 2002; Whitton et al., 2017).

Public participation is a voluntary activity, meaning that people are free to attend the process or decision-making process. Citizens are not obliged by the ruling class or powerholders to become involved in activities related to the projects and programs (Lee and Abbot, 2003; Kravchenko, 2010a). Participants can accept or support decisions freely. Individuals can become involved in public participation by submitting requests, motions, reports, petitions, and other types of application (De Santo, 2016). Another main feature of public participation is that people are not charged for participation in the decision-making process; otherwise, particularly poor communities may not afford to participate and hence be excluded from the process purely on financial grounds (Beierle, 2010). However, community representatives may demand money for the time and effort that they expend in participation; the fee depends on appropriate negotiations between representatives and those who are represented.

Actors in the Decision-Making Process

The term "public" refers to a general broad category of people because it implies all members of a society (Gunderson, 2018)—the public in China, for example, consists of approximately 1.5 billion people (Wanqing et al., 2014). It is unlikely that the entire public in a country will be engaged in a decision-making process related to the environment. However, IEL such as the Rio Declaration and Aarhus Convention provide a clearer definition of the term public—the Aarhus Convention, for example, has a separate formulation of "the public concerned," which encompasses those people most likely to be affected by, or having an interest in, the relevant environmental decision making (Mason, 2010; Lee and Abbot, 2003). However, the entire public may have a legitimate interest in, or feel concerned about, environmental decision making because major nuclear accidents can affect an entire country—or even further afield—due to the release of radioactive materials into the atmosphere.

Similarly, the public participation literature notes that those who are affected, or could potentially be affected and interested, should participate in the decision-making process (Toth, 2010; Lee and Abbot, 2003). This might not always be practicable. Firstly, the decision may affect future generations, but it is not clear how future generations can be represented in a decision-making process (Meyer, 2017). Secondly, some environmental projects, such as NPPs, are constructed on the boundary with two or more countries (Ahmedova, 2016); it is not clear how the public from in such other countries can participate in the decision-making process. The definition of the public does not include a foreign public who might potentially be affected by the decision or projects related to the environment (Gunderson, 2018; Mason, 2010).

Actors in public participation are not purely limited to those concerned citizens directly affected by decisions and government organizations. There may be more groups than lay people and powerholders participating, depending on the level of participation (Coenen, 2009). For example, according to the Aarhus Convention, "the public" means one or more natural or legal persons, and, in accordance with national legislation or practice, their associations, organizations or groups (EU, 2001). This implies that not only lay people but also nongovernmental organizations and civil society groups who have an interest or stake in an issue are also a part of the public, and can thus be involved in the decision-making process (Lee and Abbot, 2003). So, citizens can attend the decision-making process as individuals or as groups.

Types of Public Participation

There are many forms of public participation depending on the level of power shared between the public and powerholders. According to the typology defined by Arnstein (1969), there are eight possible levels of citizen participation. The bottom rungs of the ladder include manipulation and therapy, which are considered the least amount of citizen participation or nonparticipation. It is assumed that the proposed plan is the best and there is no need to change it, meaning that the first two rungs are employed only to gain additional public support for the project. It covers one-way flow of information from public institutions to the public. Manipulation is the means of participation, in that the public is manipulated into thinking that public participation is in progress (Tritter and McCallum, 2006). Those in power do not intend to incorporate public values, assumptions, and preferences into the decision making at this stage. Citizens do not have any control or decision-making role. They are used by those in power purely to demonstrate that people are participating in the decision-making process, or otherwise persuade participants to think like them in order to gain their support for the project and its planning. This participation is commonly seen in military regimes. In the case of therapy, the main aim is to "educate" or "cure" the poor and marginalized groups, in particular as a form of enlightenment enacted by mental health experts, social workers, and psychiatrists in order to bring them into line with broader society or to pathologize their attitudes about government, rather than to enable citizens to participate in planning or conducting programs. This is because powerholders assume that powerlessness is synonymous with mental illness, and that the general public is incapable of effective decision making (Carpentier, 2016). In other words, participation is merely a formality in this paradigm.

Informing, consultation, and placation occupy the middle rungs of the ladder, which are considered "tokenism." Tokenism refers to the idea that while some roles are given to the people, but they do not have any influence on any decisions (Tritter and McCallum, 2006). Informing is only the first step toward legitimate citizen participation. There is no channel for feedback, or citizens do not have any power of negotiation during informing because it requires a one-way flow of information from officials to citizens; it is a top-down initiative. Powerholders ensure that participants know what is happening or will happen in relation to the projects. However, information is shared with the public at a very late stage in the process when changes can no longer be affected by the participants. In other words, participants do not have the opportunity to influence decisions. Information can be technical, superficial, irrelevant, or incomplete beyond the understanding of lay people. Discouraging questions or giving irrelevant answers can be seen in this form of participation. The most frequent tool used for informing is the news media, pamphlets, posters, and responses to inquiries, which is one-way flow of information from powerholders to the public. The next rung on the ladder is consultation, which is one of the more common forms of participation. Citizens participate only to the extent of expressing their thoughts and concerns. Consultation provides decision makers with public values, assumptions, and preferences, but it does not offer the assurance that public concerns and ideas will be taken into account when a final decision is made. What is important to officials is how many people are involved in meetings or how many people answer survey questions, because if a lot of people participate in the decision-making process this can increase the resultant decisions' legitimacy. Consultation includes surveys, questionnaires, neighborhood meetings, public hearings, and public enquiries (Callahan, 2007). The final level of tokenism is placation, which allows selected have-nots access to advice that may facilitate the incorporation of public values into the decision-making process but maintains the right to judge the legitimacy or feasibility of the advice with those in power. One or two worthy community members might hold seats on the board, meaning that citizens, at least in this instance, might begin to gain a degree of influence. They have access to powerholders, but ultimate power still lies outside the community. For this reason, public input can be ignored when final decisions are made, particularly if their opinions are unfavorable from the perspective of the powerholders. Citizens are often not equipped and do not have enough time for proper participation (Maier, 2001).

The last three rungs at the top of the ladder include partnership, delegated power, and citizen control, which are termed citizen power and mean that control of the process is being held by the citizens themselves, either partly or completely. True and meaningful participation takes place at this stage (Enserink and Koppenjan, 2007). This form of participation gives the public a direct decision-making role. Mechanisms employ two-way flow of information to resolve impasses. In the case of partnerships, planning and decision-making responsibilities are shared between citizens and powerholders through such structures as joint policy boards or planning committees (Choguill, 1996). Power is redistributed through negotiation between officials and citizens. Stakeholders interact in order to understand each other's concerns and thoughts in this model. The partnership enables citizens to negotiate and engage in trade-offs with those in power. According to Arnstein (1969), the partnership can be the most effective form of participation if the community is well organized, meaning that they have sufficient resources to hire or fire a wide range of staff, and pay their leaders for the time they spend on participation. In the case of delegated power, citizens have a clear majority of seats on the boards that make the decisions about a particular plan or program. Citizens hold the possibility of veto over government plans. Delegated power ensures public accountability to the program or projects; it further ensures that decisions are taken or programs are implemented according to their desires. The top rung of the ladder is citizen control, where citizens hold full managerial power over a policy or plan and handle all the associated process including policy-making and management. This guarantees that citizens can govern a program or an institution and are in full charge of policy and managerial aspects with no intermediaries between themselves and the source of funding. It is also the least common type of participation. Citizen control requires have-not citizens who are willing to attend activities and spend time and effort in participation (Tritter and McCallum, 2006).

Access to Justice

Broadly speaking, access to justice refers to the idea that citizens have a right to participate in the justice system in the event that a decision is reached that will potentially impact on their lives (De Sadeleer et al., 2005). More specifically, justice carries four main implications (Anderson and Boyle, 1996). Firstly, it implies the possibility of judicial review with respect to environmental decision making (Armeni, 2016; Oliver, 2013). People have a right to bring enforcement proceedings to environmental decisions in an administrative review or other independent and impartial body before the courts in cases of a decision which has affected, or will potentially affect, their lives and the environment (May, 2013; Wates, 2005). Members of the public can also challenge the substantive and procedural legality of environmental decision making and be allowed a review procedure before a court of law or other (De Sadeleer et al., 2005). The benefits and impacts of projects or decisions are reviewed by an independent and impartial body established by law, which is independent of the government that makes/formulates the environmental decisions, regulations, or policy (Millner, 2011). The main rationale here is to save time and money because a court review can be lengthy and expensive, and environmental information and decisions might be needed quickly (Pirro, 2008). Additionally, any individual who considers that his or her request for information was not addressed in accordance with national law can have access to a judicial or nonjudicial review procedure (Rose-Ackerman and Halpaap, 2002).

Secondly, access to justice includes the right to judicial redress with respect to the impact of environmental decisions, projects, policy, and so on, on people or the environment (Hajjar Leib, 2017; Oliver, 2013). Environmental harm is not shared equally among individuals in a society (Bezdek et al., 1995). Particularly disadvantaged or marginalized groups bear the brunt of the environmental impacts of human activity because the environmental risk is not distributed fairly throughout society (Schlosberg, 2009). For example, while people who live within 30 km of the damaged power stations in Fukushima do not have access to their property, other Fukushima residents do and are still able to enjoy their property despite the FNA (Millner, 2011). This shows that the environmental impacts of the FNA may not be "shared" equally, or do not affect everybody to the same extent, depending on the level of environmental harm. Access to justice enables these groups to seek redress and compensation for the damage to their lives, health, property, and so on, as guaranteed by the constitution (Wates, 2005; Armeni, 2016; Oliver, 2013). When the right to life or health is violated because of ecological contamination, the affected individual, or individuals, may need to seek remedy for such violations.

Thirdly, access to justice indicates that the justice system has to be equally accessible to all individuals of a society (Rose-Ackerman and Halpaafsap, 2002; Freitas, 2017). Access to justice enables members of the public to access the law, a fair, well-known and effective legal order, with access to the courts or alternative mechanisms (Anderson and Boyle, 1996). However, the recognition of the access to justice by law is not enough in itself to allow it to be enjoyed by individuals because there are some inherent barriers to such access (De Sadeleer et al., 2005; Veinla, 2005). The first of these is geographical barriers (May, 2013; Wates, 2005). Generally speaking, courts and tribunals that make decisions on matters of environmental law tend to be based in the capital cities (Anton and Shelton, 2011; Pirro, 2008). It might thus be expensive and time consuming to get access to courts and tribunals from particularly rural areas (De Sadeleer et al., 2005; Veinla, 2005). Secondly, social awareness is key to achieving access to justice because if people are not sufficiently aware of their rights or are not educated about how they may access tribunals and courts and their procedures, enjoying their right to access to justice can be difficult in practice (Anton and Shelton, 2011; Boyd, 2011). Citizens, however, can get assistance from lawyers and experts but this can be expensive, particularly poor community groups (Hajjar Leib, 2017; Oliver, 2013). It may not be possible for individuals to meet the costs of obtaining these services (Armeni, 2016). However, this issue can be mitigated through legal aid which provides assistance to people who are unable to afford legal representation and access to court systems (May, 2013; Wates, 2005). Thirdly, the high cost of litigation, especially the associated legal fees, can themselves be an obstacle to the enjoyment of access to justice by poor communities (Wates, 2005; Armeni, 2016; Oliver, 2013). The cost depends on the quality of the applicant's case (Pirro, 2008). Fourthly, if the person or NGOs lose their lawsuit against the third party or the state, he or she has to pay the costs of the other side (Humby, 2010). Plaintiffs or submitters may have little to gain financially from winning a lawsuit, but merely bringing a lawsuit can of itself represent an unacceptable financial risk. These are some obstacles, both material and psychological, for the general population to exercise their access to justice.

Fourthly, access to justice must have socially fair results, namely, through judicial independence from other branches of government or private and partisan interests (Rhode, 2004). Courts and judges must not be subject to improper influence from other actors, whether governmental or private (Anton and Shelton, 2011). Judges must be able to perform their duties free of any influence or control by other actors, and must be politically insulated from the legislative and the executive powers; otherwise, courts may rule in favor of those in power without making rulings in proper accordance with the rule of law and judicial discretion (Rhode, 2004). Indeed, this independence is guaranteed by law in modern democracies. Article 76 of the Constitution of Japan, for example, states: "All judges shall be independent in the exercise of their conscience and shall be bound only by this Constitution and the laws" (KANTEI, 2018). This might be particularly important in the human rights framework because human rights are generally violated by the various branches of government (Boyd, 2012).

CONCLUSION

This chapter has focused on three main topics: it first discussed the relationship between human rights and the environment by indicating what a safe environment actually might be, particularly with regards to the study of nuclear energy. It has concluded that there is no such thing as zero risk but that radiation levels below 4 mSv are seen as an acceptable safety level by international institutions and countries with NPPs. However, environmental safety remains a subjective judgment because environmental risks and benefits can be perceived differently depending on one's values. Accordingly, there seems to be the need to determine what sociably acceptable safety levels actually are. Another result is that a safe environment is essential to the enjoyment and realization of human rights because people are dependent on the environment to meet their basic needs, including water, air, and food, for survival. Ongoing environmental issues such as climate change clearly demonstrate that the quality of the environment affects people's lives and health, which are recognized as human rights.

Second, we have argued that there are three distinct human rights approaches to environmental safety, which offer different challenges and opportunities to achieving a safe environment. First of all, some scholars argue that the right to the environment should be recognized as a human right because it may oblige countries to achieve environmental safety, as states have a positive duty to protect human rights. There is, however, uncertainty about how a safe environment can be achieved as the world is already polluted. In contrast, some argue that there is no need for a distinct right to the environment because human rights, including the rights to life and health, already require a safe environment—this is referred to as a reinterpretation of human rights. This has two main advantages—similar to the right to environment, it may oblige countries to preserve nature in the same way that states are currently obliged to protect human rights. It may also enable citizens to receive redress for human rights violations that result from any contamination of the environment. Another group argue that Procedural Human Rights (PHRs) can be used by concerned citizen to achieve environmental safety as they have various associated advantages, including enabling powerholders to incorporate different groups' values, risk perception, and concerns into decisions; it enables low-income communities to raise their concerns with powerholders; it may allow transparency to be achieved in environmental governance; and socially acceptable environmental safety and risks can be determined through PRs. PRs also carry certain disadvantages, in that they can be costly and time consuming, as depending on the type of public participation.

Third, we argued that SHRs and PHRs are not alternatives to each other, a fact that is also recognized by international courts such as the European Court of Human Rights. Substantive rights and PRs compete each other in terms of achieving environmental safety: one may be ineffective in the absence of the other. For example, if a person is unable to enjoy their right to health for environmental reasons, he or she may not participate in decision-making processes due to health issues. Similarly, if he or she does not enjoy the right to life due to contamination, he or she may need to apply to the courts. These examples mean that substantive rights and PRs can be effective only when they are mutually enjoyed by individuals.

The empirical chapters investigate some of these issues within the context of the Fukushima accident. It first analyzes Japan's nuclear and human rights policies to explore the cross-cutting issues between the two policy domains, as nuclear energy has affected the enjoyment of human rights in Fukushima. Then chapter 4 focuses on SHRs and the link between the quality of the environment and the enjoyment of SHRs, including the right to life, health, and property. The chapter investigates the environmental impact of the FNA on people's right to life, health, and property in Fukushima. Chapter 5 investigates how the residents have experienced the three PHRs, including the right to public participation, information, and justice to achieve environmental safety and influence Japan's energy policy pre- and post-Fukushima.

Chapter 2

Japan's Nuclear Energy and Human Rights Policies

INTRODUCTION

The link between human rights and the environment has been discussed since the 1972 Stockholm Declaration (Anton and Shelton, 2011). A clean/safe environment is seen as integral to the full realization of a wide range of human rights, including the right to life, health, food, property, and so on, by legally binding regional human rights treaties which include the African Charter on Human and Peoples' Rights and the American Convention on Human Rights Pact of San José; however, there is no recognition of such at the international level (Anderson and Boyle, 1996). While some countries (such as Turkey and Colombia) explicitly recognize the relationship between human rights and the environment within their constitutions or human rights acts, other countries (such as Japan and China) do not reference this relationship in any way (Xiao et al., 2017; Boyd, 2012; Hancock, 2003).

The link between human rights and the environment seems particularly evident in the Fukushima matter because the FNA contaminated the environment that thousands of people still, in fact, live in (Bunn and Heinonen, 2011). Japan is an established democracy with a constitution that guarantees human rights (Fujioka, 2003) and has recognized universal human rights such as the right to life, health, education, and so on (KANTEI, 2018); however, the enjoyment and realization of universal human rights are currently at risk due to the environmental contamination caused by the FNA. Accordingly, environmental safety appears to be a cross-cutting issue for Japan's human rights and nuclear energy policies as nuclear accidents pose a risk to the environment that people are dependent on for the enjoyment of human rights, according to EHRs theory (Anderson and Boyle, 1996). However, how these two policy domains are related to each other has not been explored in the literature to date.

In order to better understand the cross-cutting issue between Japan's HRP and NEP, this section aims to analyze key policy arguments, main objectives, and the issues underlying Japan's HRP and NEP. The main purpose of this section is to analyze how Japan's HRP and NEP relate to each other and consists of three main parts. The first analyzes Japan's HRP in connection with the IHR law and answers the question of how Japan's HRP addresses EHRs including civil and political rights, PHRs, and the right to the environment. Second, it analyzes Japan's NEP pre- and post-Fukushima. By doing so, we will better understand Japan's approach to the EHRs, and the connection between Japan's NEP and HRP.

JAPAN'S HRP AND EHRS

Overview of Japan's HRP

Lack of Human Rights Act in Japan

Japan does not have any human rights act at the national level; rather, the country guarantees human rights within its constitution, which is the supreme law of Japan (Port, 1991). Article 11 of the constitution states: "*The people shall not be prevented from enjoying any of the fundamental human rights. These fundamental human rights guaranteed to the people by this Constitution shall be conferred upon the people of this and future generations*" (KANTEI, 2018). Article 11 indicates that universal human rights are one of the primary principles of the Japanese state. The article, however, is somewhat vague in the sense that it does not mention what kinds of human rights the constitution actually refers to (Maki, 1990). Human rights may vary from country to country as a function of historical basis, culture, religion, tradition, political system, and so on (Goodale et al., 2007); for example, while the right to environment is recognized as a basic human right by fifty-six countries in their constitutions, other countries do not allow such recognition (Boyd, 2011). Nevertheless, as Japan does not have any human rights act at the national level, one can interpret this to mean that human rights in Constitution of Japan refer to the ones granted in the IHR documents (González, 2007).

More specifically, the term "fundamental human rights" refers to universal human rights recognized by the UDHR, which is a fundamental constitutive human rights document of the United Nations (Nakanishi, 2017). The IHR documents that Japan has ratified (such as ICESCR and ICCPR) reference the respect and recognition of human rights that UDHR guarantees (Neary, 2002). The fundamental human rights in Constitution of Japan, therefore, refer to the universal human rights addressed by the UDHR including the right to life, health, education, work, ownership of property, liberty and security of person, religion, food, water, trial, association, assembly, expression, privacy, free movement, asylum, marriage and family, social security, equality, freedom from discrimination, freedom from slavery, freedom from inhumane treatment, legal recognition, equality before the law, remedy by competent tribunal, freedom from arbitrary legal prosecution, fair public hearing, marriage, participate in government, rest, justice, adequate living standard, participate in and enjoy the culture of one's community (UN, 1948).

Japan has chosen not to have a national human rights act which declares certain basic human rights that Japan is legally required to guarantee. A human rights act is a law that allows citizens defend their rights in the courts and compels public organizations such as the government, police, and local councils according to the related legislature (Kavanagh, 2009). If a public authority breaches a person's human right, he or she may be able to take action under the act (Bellamy, 2011). In order to protect everybody's human rights, some countries have their own national human rights acts, for example, the United Kingdom, Canada, Australia, New Zealand, the United States, and so on (Donnelly, 2002). A clear, informative, and well-written human rights act may contribute to increasing citizens' awareness about human rights in Japan. This could also guide the Japanese courts when they address any lawsuits in relation to human rights issues. Additionally, the act contains more detailed provisions than the constitution. While the constitution establishes the main principles with regards to human rights, the Human Rights Act gives more information about the concept and scope of these rights, which can thus better guide individuals and public institutions (Vick, 2002). The constitution defines the basic pattern of authority, not only human rights, which makes it more complicated than a single human rights act which, of course, declare only human rights (Bellamy, 2011). Therefore, the lack of national human right act appears to be a weakness in Japan's HRP.

IHR Treaties That Japan Is Party to

The fact that Japan guarantees universal human rights is emphasized by several HRP documents which set out the Ministry of Justice's intent to achieve (Maki, 1990). "The National Report by Japan for the Universal Periodic Review" which was published in 2008, for example, states: "*Japan recognizes the realization of the universal values common to the international community, such as fundamental human rights*" (MOFA, 2008). According to the report, Japan's HRP is based on three principles including (1) adoption of universal human rights into national law, (2) raising people's awareness about human rights and providing consultations on human rights, and (3) protection of universal human rights in criminal justice proceedings. Similarly, another HRP document, entitled: "Japan's Human Rights Commitments and Pledges" declares that promotion of human rights is a legitimate interest of the country, and protecting universal human rights is the most fundamental responsibility to face Japan's government. More specifically, the report states that Japan is committed to the implementation of universal human rights guaranteed by IHR documents to which Japan is signatory (MOFA, 2013). These include:

The International Covenant on Economic, Social and Cultural Rights (ICESCR) (1979)
The International Covenant on Civil and Political Rights (ICCPR) (1979)
The Convention on the Elimination of All Forms of Discrimination against Women (CEDAW) (1985)
The Convention on the Rights of the Child (1994)
The International Convention on the Elimination of All Forms of Racial Discrimination (1995)
The Convention against Torture and Other Cruel, Inhuman or Degrading Treatment or Punishment (1999)
The International Convention for the Protection of All Persons from Enforced Disappearance (2009)
The Convention on the Rights of Persons with Disabilities (2014)

The above human rights documents target different groups in any given society such as the rights of disabled people and the rights of immigrants. This means that Japan's HRP is not limited purely to the adoption of fundamental human rights. The country also intends to promote such diverse human rights which are not referenced by UDHR, for specific groups such as children, women, indigenous people, and so on. Japan's HRP is, therefore, based on promotion and protection of not only fundamental human rights but also specific human rights that address the needs of diverse groups in its society (Maki, 1990; Port, 1991).

Human Rights Documents Ratified by Japan

Japan has ratified only three of the above documents: ICESCR, ICCPR, and CEDAW (Port, 1991). ICESCR was adopted by the United Nations General Assembly in 1966 and ratified by Japan in 1978 (Maki, 1990). The treaty commits its parties to work toward granting several human rights including the right to life, to health, to property, to education, to work, and so on (Beer, 2009). Another document that Japan has ratified is the ICCPR, which was adopted by the United Nations General Assembly in 1966 (Iwasawa, 1998). ICCPR obligates countries that have ratified the treaty to protect and preserve basic human rights at the level of their national law in areas such as the right to life, freedom of speech, and freedom of assembly (Steiner et al., 2008). Both ICESCR and ICCPR compel parties to take administrative, judicial, and legislative measures to protect the rights enshrined in the treaty and to provide effective legal remedies (Viljoen, 2007). When human rights are violated, the treaties require remedies for the victims (Shelton, 2006).

In theory, the IHR treaties that Japan has ratified have the same force as domestic law (Fujioka, 2003; Port, 1991). Article 98 (2) of Constitution of Japan states that "*the treaties concluded by Japan and established laws of nations shall be faithfully observed*" (KANTEI, 2018). This is also underlined by Japan's basic human rights documents—the latest policy document which declares its national commitments and pledges for the promotion and protection of human rights, entitled "*Candidate for the Human Rights Council 2017–2018*" which, for example, states that "*Japan honestly has implemented the obligations of international human rights instruments that Japan has concluded*" (MOFA, 2017). This implies that Japan has adopted the main principles of the IHR documents that the country has previously ratified.

In practice, IHR treaties have two main implications for Japan's legal norms. Firstly, the IHR law profoundly influences the development of domestic law (Port, 1991). The National Diet (Japan's bicameral legislature) has actively integrated international legal obligations such as ICESCR and ICCPR into Japan's domestic legal framework through the revision of old laws or the creation of new ones (Iida, 2004). As an example, after ratifying the Women's Convention in 1985, the Japanese legislature amended the Nationality Act such that both mother and father confer nationality on their children—prior to the Women's Convention, only Japanese fathers could pass their citizenship on to their children; subsequently, however, Japan, as a state party, was obliged to realize the equality between women and men within its national borders (Alston, 1999). Another example is that Constitution of Japan guarantees a number of civil and political rights such as right to expression, right to association (Article 21), and right to petition (Article 16), which are the main principles of the International Covenant on Civil and Political Rights, to which Japan is also party (Webster, 2009).

Secondly, IHR treaties have a self-executing effect on domestic law (Steiner et al., 2008). The Japanese Supreme Court increasingly interpret and apply international law in human rights litigation without recourse to any legislative provision, particularly in cases involving criminal procedure and minority rights when the provisions of the civil code do not provide a solution to the issue or domestic law is inconsistent with IHR treaties (Webster, 2009). For example, a Nigerian defendant required the services of an interpreter in 1993 and the trial court, pursuant to domestic law, charged the convicted defendant court costs for interpretation fees, which went directly against the ICCPR because the treaty guarantees the free assistance of an interpreter. The trial court deferred to domestic statutory law and the Tokyo High Court ordered to provide free interpretation to criminal defendants by giving effect to the ICCPR's guarantee of free interpretation services (Webster, 2009). This example shows that Japanese judges tend to domesticate IHR treaty obligations.

To sum up, Japan is generally quite diligent in the implementation of human rights treaties to which it is a signatory. The country has guaranteed a number of internationally recognized human rights in its constitution such as the right to life and to health.

Japan's HRP on EHRs

The above analysis shows that Japan has a tendency to adopt IHR treaty obligations. The IHR, however, evolves in such a way that EHRs become core human rights within the IHR framework, such as in the European Court of Human Rights, the Inter-American Commission of Human Rights, and the African Commission on Human and Peoples' Rights (Anton and Shelton, 2011). This section analyzes how Japan recognizes EHRs within its legal framework. This section aims to analyze Japan's policy on human rights in connection with the environment.

The Lack of Human Rights in Connection with the Environment in Constitution of Japan

The reinterpretation of human rights means that a safe environment is already embodied within the penumbra of internationally established human rights such as the right to life and health, property, and so on (Anton and Shelton, 2011). The Constitution of Japan has 103 articles which reference a number of universal human rights, and which form the core HRP in Japan. These include: (1) the right to life—Article 13 states: "*We recognize that all peoples of the world have the right to live in peace, free from fear and want*"; (2) right to health—Article 25 states: "*All people shall have the right to maintain the minimum standards of wholesome and cultured living. In all spheres of life, the State shall use its endeavours for the promotion and extension of . . . public health*"; (3) right to property—Article 29 states: "*The right to own or to hold property is inviolable*"; (4) right to education—Article 26 states: "*All people shall have the right to receive an equal education*"; and (5) right to work—Article 27 states: "*All people shall have the right and the obligation to work*" (KANTEI, 2018). All these human rights are recognized by the UDHR and ICESCR that Japan is a party to (Port, 1991; Maki, 1990). Therefore, Japan guarantees several of the universal human rights recognized by IHR within its own national constitution.

However, this recognition of universal human rights does not mean that a safe environment is necessarily *guaranteed* by the Constitution of Japan.

From an EHRs perspective, these rights require a safe environment that can be enjoyed and realized by individuals because the environment provides the basic human needs such as the air, water, and food which are fundamental to survival (Anton and Shelton, 2011). Japan, however, does not have any policy or legal framework which implies or declares that human rights require a safe environment. The reinterpretation of human rights is relative and based on the interpretation that a safe environment is considered essential to the enjoyment of human rights by EHRs theory (Xiao et al., 2017). This interpretation may be not the case for some, or indeed many, constitutional rights. For example, Article 20 in Constitution of Japan states: "*Freedom of religion is guaranteed to all*" (KANTEI, 2018). It seems reasonable to claim that a clean environment is not a precondition to the realization of the right to religion, and nor is there any direct link between a safe environment and freedom of religion. For this reason, it seems essential to determine and declare which human rights require a safe environment by a well-defined HRP, and which in Japan does not currently exist. Therefore, Japan has guaranteed human rights in the constitution, but has not set policy as to whether human rights require a safe environment.

No Recognition of Right to the Environment in Japan

Right to the environment means that everybody has the right to a general satisfactory environment favorable to their development (Thorme, 1990). Fifty-six countries currently recognize the right to the environment as human rights within their constitutions (Boyd, 2011). However, Japan itself does not have any HRP related to the distinct right to the environment; neither the Constitution of Japan nor any other legal documents in Japan reference any human rights in connection with the environment. Japan tends to adopt internationally recognized human rights such as the right to life and health, but IHR documents that Japan has ratified to date do not *explicitly* guarantee the right to environment. Furthermore, there is no one universal human rights document which recognizes this right as a human right in a manner that could inspire Japan to make such an adoption (Anton and Shelton, 2011).

Civil and Political Rights

Civil and political rights refer to rights that protect people's freedom from infringement by governments, social organizations, and private individuals (Piovesan, 2004). Japan guarantees several civil and political rights in its constitution, which include: (1) right to assembly, to the association and to expression—Article 21 states: "*Freedom of assembly and association as well as speech, press and all other forms of expression are guaranteed*"; (2) right to freedom of thought—Article 18 states: "*Everyone shall have the right to freedom of thought*"; (3) right to petition—Article 16 states: "*Every person shall have the right of peaceful petition*" (KANTEI, 2018). The right to assembly in Article 21 is linked with the right to protest because it applies to marches and demonstrations, press conferences, and public and private meetings, among others. These articles indicate that civil and political rights are well established within the country.

Japan aims to develop policies for the promotion and protection of human rights and fundamental freedoms in order to meet universal human rights standards, which is clearly underlined by reports on human rights released by the Ministry of Justice and Ministry of Foreign Affairs. One, for example, stated that "*Protecting human rights is the most fundamental responsibility*" (MOFA, 2009, p. 1). Indeed, the country ratified the ICCPR in 1979, which is a key IHR treaty, providing a range of protections for civil and political rights (Maki, 1990; Port, 1991). The ICCPR obliges state parties to guarantee a various range of civil and political rights, which are all adopted by Constitution of Japan (Piovesan, 2004).

Access to the Information

Access to information is a well-established human right in the IHR documents, such as CEDAW and ICCPR, that Japan has ratified (Neary, 2002). Article 19 of the ICCPR, for example, states: "*The Everyone shall have the right to freedom of expression; this right shall include freedom to seek, receive and impart information*" (OHCHR, 1976). Japan does not explicitly recognize the right to receive and impart information within its constitution and none of the HRP documents references this right. The Supreme Court has, however, decided that the right to information is protected by Article 21 of the constitution which guarantees the right to assembly, to the association, and to expression because the people need to receive and impart information to enjoy Article 21 (Boling, 1998). The Supreme Court's decision seems important for two reasons. Firstly, the right to information is guaranteed by Constitution of Japan. Receiving information might be seen as essential for the realization of civil and political rights because if people do not have sufficient information about a given event, policy, decisions, issues, and so on, they may not be able to have, express, share their opinions about it (Anton and Shelton, 2011). Secondly, how the Supreme Court in Japan interprets Article 21 in Constitution of Japan, which has decided that right to information is embedded in the rights to assembly, to association and expression is very similar to Article 19 of the ICCPR. In other words, this is evidence that the CCPRR influences not only the HRP of the country but also judicial decisions related to human rights in Japan.

Public Participation in Decision-Making Processes

Public participation in decision-making processes is a human right that was first recognized by the UDHR in 1948 (Boyle, 2007), and which Japan is a state party to (Port, 1991). Many of the legally binding regional human rights documents also protect public participation in decision-making processes such as the European Court of Human Rights, the Inter-American Commission of Human Rights, and the African Commission on Human and Peoples' Rights (Anton and Shelton, 2011); similarly, the ICCPR that Japan has ratified guarantees public participation in decision-making processes. Article 25 states: "*Every citizen shall have the right* [sic] *to take part in the conduct of public affairs, directly or through freely chosen representatives*" (OHCHR, 1976). The Supreme Court in South Africa, for example, took the view that this article in the ICCPR imposes an obligation on state parties to take positive steps to ensure that their citizens have the opportunity to be involved in decision-making processes (Syma Czapanskiy and Manjoo, 2008). This is, however, not recognized or guaranteed by Constitution of Japan in any way; indeed, Japan does not have any explicit HRP on public participation in decision-making processes.

Access to the Justice

Access to the courts is a human right recognized by various human rights documents (Oliver, 2013; Beer, 2009). The ICCPR, as ratified by Japan, obliges states to take positive steps to ensure that their citizens have appropriate opportunity to access the courts to seek remedy for any violation of human rights guaranteed by the appropriate constitution or the ICCPR (Piovesan, 2004). Japan expressly guarantees access to the courts in its constitution. Article 32 of the Constitution of Japan states: "*No person shall be denied the right of access to the courts*" (KANTEI, 2018). This means that Japanese citizens explicitly have this right. Article 76 states: "*All judges shall be independent in the exercise of their conscience and shall be bound only by this Constitution and the laws*" (KANTEI, 2018). This seems important because human rights are generally violated by the public authorities (Maki, 1990); the principles, however, state that Japan's judiciary is independent of any person or institution, including the government. Additionally, Article 17 states "*Every person may sue for redress as provided by law from the State*" (KANTEI, 2018). These three human rights mean that victims of the FNA have the right to apply to the courts to sue the Japanese state or TEPCO for damages to their lives, health, property, and other affected human rights such as the right to education. The right to access to justice is, therefore, a well-established right within the Constitution of Japan.

JAPAN'S NEP

The above sections analyzed Japan's HRP. As this chapter aims to analyze the connection between Japan's human rights and its energy policy, the following section will analyze Japan's NEP and how it has evolved since 2011 since the FNA. Accordingly, this section is divided into three parts: it first analyzes nuclear energy regulation, then Japan's energy policy before the accident, and finally key policy changes in the aftermath of the accident.

Overview of Nuclear Energy Regulation in Japan

International Nuclear Law

Nuclear energy law (NEL) is a nuclear energy legislation related to the peaceful use of nuclear science and technology. It is a paradigmatic example of multipolar law, whereby national, regional, and international law govern this field (Vadi, 2016). NEL addresses, in the main, issues of liability and compensation for nuclear-related damage, international trade in nuclear materials and equipment, and also serves as a center for nuclear law information and education (OECD, 2006). It also focuses on establishing legal mechanisms to both prevent and mitigate the risks inherent to the use of nuclear energy (Pelzer, 2009). NEL's remit, therefore, is to achieve higher levels of safety for all countries with NPPs. Japan itself is a party to the following conventions relating to nuclear safety:

- Convention on Nuclear Safety
- Joint Convention on the Safety of Spent Fuel Management and on the Safety of Radioactive Waste Management
- Convention on Early Notification of a Nuclear Accident
- Convention on Assistance in the Case of a Nuclear Accident or Radiological Emergency

Japan's main energy policy is to fulfill its obligations as described by the Convention on Nuclear Safety (NRA, 2016). According to the Nuclear Regulation Authority (NRA), the Convention on Nuclear Safety has the same legally binding power as domestic laws (Kingston, 2013a). The Convention on Nuclear Safety was adopted by IAEA in Vienna on June 17, 1994 (Domaratzki, 1995), and was the world's first binding international standard on the siting, construction, and operation of land-based nuclear facilities (Kamminga, 1995). It imposes a set of obligations on its seventy-eight state parties (including Japan) to implement certain safety rules at all civil facilities related to the use of civilian NPPs (Johnson, 2013). These include design and construction; issues of site selection; operation and safety verification;

and, lastly, emergency management and preparedness. State parties are obligated to adopt the Convention's standards within their national regulations (Domaratzki, 1995).

The overriding problem, however, seems the lack of any effective enforcement mechanism, despite being hard law (Jankowitsch and Tonhauser, 1997). The only responsibility of the contracting parties is to submit a report presenting the measures taken to implement the obligations of the Convention in review meetings that are held every three years (Domaratzki, 1995). The reports are only discussed with other state parties, after which any progress achieved, and any existent difficulties, are discussed (Kamminga, 1995). This may be, arguably, interpreted in a way that the Convention implies that responsibility for nuclear safety is a domestic, rather than international, matter, which means that nuclear safety ultimately falls under the exclusive responsibility of the state (Montjoie, 2016); the Convention itself does not deem to interfere in the competence of national regulatory bodies (Jankowitsch and Tonhauser, 1997). However, nuclear accidents affect countries other than the ones in which they occur significantly. The Chernobyl plant, for example, was located in Ukraine but had a major impact on other countries including Russia, Belarus, and Turkey among various other European states. From this perspective, an enforcement mechanism enacted at the international level would seem vital to the control safety standards of countries with NPPs.

Nuclear Legislation in Japan

Nuclear reactors are regulated by the Atomic Energy Basic Law (AEBL), the Reactor Regulation Act (RRA), the Radiation Hazards Prevention Act (RHPA), and the Nuclear Emergency Act (NEA), which are all legally binding (OECD and NEA, 2006). The AEBL was enacted in 1955 to regulate the research, development, and utilization of atomic energy (AEC, 2018). It establishes the framework for the regulation of nuclear activities. The AEBL ensures the protection of the people's lives, health, and property, and the environment in nuclear energy-related activities (AEC, 2018).

The RRA, as enacted in 1957, provides the necessary, comprehensive regulations for the use of nuclear energy including waste disposal, use of nuclear fuel materials, and so on, as well as on the installation and operation, and so on, of reactors (OECD, 2017). It obliges operators to be responsible for establishing appropriate security rules and procedures in terms of the construction and operation of nuclear facilities and ensure public safety by accounting for the hazards that might arise from nuclear energy-related materials and reactors. It also sets out essential rules that must be observed in international conventions such as the Convention on Nuclear Safety (OECD, 2017).

The RHPA Law governs radiation protection issues by regulating the use, disposal, and sale of any radionuclides and ionizing radiation-generating equipment (NSR, 2018). The Prevention Law also sets out dose limits for exposure to ionizing radiation, which is 50 mSv/yr for workers and 1 mSv/yr for members of the public (NSR, 2018). This limit has been changed by the government such that the dose limit is 100 mSv/yr for workers and 20 mSv/yr for private individuals after the FNA accident. There is, however, no scientific explanation as to why the public dose limit is 20 mSv. It is also not known why the government changed the public dose limit from 1 to 20 mSv (Kingston, 2013b). As a radiation dose above 1 mSv before the FNA was not safe, it is not particularly clear how a radiation dose of 20 mSv can be safe (NPA, 2018). This seems a political decision rather than a scientific one (Kawano et al., 2012), as there is no clear scientific explanation for this change.

Japan has established the NEA for emergency preparation in the instance of a nuclear incident or accident (OECD, 2017). The main objective of the NEA is to protect citizens' lives, health, and property from nuclear incidents and accidents by strengthening the response measures needed to address nuclear emergencies (NRA, 2016). The NEA obliges both companies and the government to take all possible steps to prevent a nuclear event, prevent the spread of the effects of a crisis, and repair any damage caused by such an incident or accident by strengthening response measures to nuclear emergencies (OECD, 2017).

Japan's NEP before the FNA

Japan is the third largest consumer of nuclear energy worldwide, with fifty-one light-water nuclear power reactors operated by ten electric power companies (Kingston, 2013a). Nuclear energy was a major pillar in Japan's longer-term energy policy, particularly since the 1973 oil crisis—oil was Japan's principal source of fuel for the generation of electricity, with NPPs generating only 2.6% of Japan's electrical power until this crisis (Hayashi and Hughes, 2013). The country was generating roughly 30% of its electrical power from NPPs, with a total installed generating capacity of around 49 GW, at the time of the Fukushima accident (Shadrina, 2012). Although Japan's Basic Energy Plan, which was adopted by the Cabinet in 2002 and revised in 2010, called for an increase in Japan's reliance on nuclear energy by building nine new nuclear reactors by 2020 and a total of fourteen additional plants by 2030 (Hayashi and Larry, 2013), the country's NEP is now under discussion due to the FNA (Suzuki, 2015).

Japan's NEP has been predominantly shaped by the pronuclear Liberal Democratic Party of Japan (LDP), which has been the prime mover behind

NEP in Japan for fifty years (Watanabe, 2016). Japan's NEP before the accident had been formulated to achieve three fundamental goals (Komiyama and Fujii, 2017). The first of these was energy security which refers to a stable and domestic supply at affordable prices. The country's self-sufficiency rate of primary energy sources excluding nuclear energy is 6% (JAEC, 2015). Japan is the fifth largest energy consumer worldwide, but as a nation has few natural mineral resources (Shadrina, 2012). The country relies on imports for almost all of its oil, natural gas, and coal. It ranks fifth worldwide in terms of petroleum imports, and first for liquefied natural gas (LNG) (Koyama, 2013), meaning that a high dependence on imported fossil fuels endangers energy security in Japan. The second goal is economic efficiency—this implies the realization of low-cost energy supply by enhancing efficiency. While 1 kW h costs roughly 5 yen in nuclear energy, it costs 6 yen in coal, 7 yen in LNG, 10 yen in hydroelectric, 13 yen in oil, 14 yen in wind, and 46 yen in solar energy, respectively, none of which include environmental costs (Hong et al., 2013; Kingston, 2013b). Based on this, nuclear energy emerges as the optimal source of energy by which to achieve economic efficiency (Duffield and Woodall, 2011).

The third goal is environmental sustainability, which particularly targets the pursuit of environmental protection in terms of climate change mitigation (Duffield and Woodall, 2011). Japan's NEP documents regard nuclear energy as an effective way to combat climate change (JAEC, 2017). Japan is aiming to reduce its CO_2 emissions by 25% by 2020 (as compared to 1990 levels) and by 26% by 2030 compared to 2013 levels in accordance with its legally binding obligations under the Kyoto Protocol and the 2015 Paris Agreement, respectively (Komiyama and Fujii, 2017). The key to achieving these targets was identified as being the increased use of nuclear power, as NPPs are viewed as a rather ecologically clean option due to their low greenhouse emissions (Lenzen, 2008). The three are often collectively referred to as the Energy Security+ Economic Efficiency+ Environmental Sustainability "3Es" policy (Shadrina, 2012). Therefore, NPPs were an important baseload power source through which Japan planned to achieve the "3Es."

On the other hand, the disadvantages of nuclear energy may outweigh its advantages in terms of achieving the 3Es policy. Firstly, nuclear energy creates its own environmental issues including: (1) it releases gaseous and liquid radiological effluents into the environment; (2) it produces nuclear waste (Cotton, 2009); (3) the cooling systems destroys marine organisms, such as fish, seals, shellfish, and turtles; (4) uranium mining affects surface water quality and quantity, groundwater quality and quantity, soil, air quality, and biota which are damaging effects of nuclear energy on the environment (McCombie and Jefferson, 2016; Akyuz, 2017). Secondly, nuclear energy seems the cheapest energy source with which to achieve economic efficiency in Japan, but this does not include its environmental costs such as nuclear waste and nuclear incidents/accidents (McCombie and Jefferson, 2016). All NPPs produce high-level radioactive wastes (HLW) which are stored at surface interim storage facilities in Japan, but this is a huge risk, and indeed a burden, for future generation to manage HLW on the ground over tens of thousands of years (Crowley, 1997). Additionally, NPPs are vulnerable to accidents, natural events, and terrorist attacks which is another risk to the environment and society, for example, it is estimated that the decontamination process from the Fukushima disaster will cost around $20 bn (£14 bn) (Starr, 2016). The cost of how the accident has affected residents and animal life, and how it will continue to affect future generations cannot be measured (Kingston, 2013a). Therefore, nuclear energy might not be cheap due to its environmental cost.

Japan's NEP after the Fukushima Accident

The Fukushima disaster triggered major reforms in Japan's energy policy (Koyama, 2013; Cotton, 2015). Historically speaking, Japan's NEP has been amended several times due to national and international crises including nuclear accidents (e.g., the Tokaimura nuclear accident), the economic crisis (such as the two oil crises in the 1970s), institutional stasis, and so on (Shadrina, 2012). Each crisis or national tragedy can trigger nonincremental or major policy changes (Bennett and Howlett, 1992). It seems to be the case that the Fukushima disaster has indeed become a wake-up call for a new NEP—Tanaka (2013, p. 243) called this turning point the "Big Bang in Japan's energy policy"—because the public trust in NPPs has decreased to 20% support due to administrative, regulatory, and safety failings in/during the accident (Fam et al., 2014; Suzuki, 2015). Accordingly, Japan has made the following reforms in its energy policy as a response to the FNA.

Safety as a Key Principle

Japan's current NEP is principally based on the Basic Energy Plan developed under the Basic Act on Energy Policy, which entered into force in 2002 and was last revised in 2017 (Suzuki, 2015), and the Fourth Strategic Energy Plan documents, which were adopted by the government in 2014 (IAE, 2016). Both documents advocate restarting NPPs by providing the basic principles of Japan's NEP, which aims to maintain a balance between safety/security and energy security (stable supply), economic efficiency (cost reduction), and the environment (environmental sustainability). This represents a new NEP in Japan, referred to as the Energy Security+ Economic Efficiency+ Environmental Sustainability+ Safety "3Es plus S" policy, which prioritizes safety with the aim of preventing such disastrous accidents in the future (Vivoda and Graetz, 2015).

After the FNA, the Atomic Energy Basic Act was revised in 2012, and a new objective of securing nuclear safety was added in Article 2, which states: "*The research, development and utilization of nuclear energy shall be limited to peaceful purposes, shall aim at ensuring safety. If the licensee violates this order, the NRA has a right to shut down the reactor*" (IAEA, 2018). According to the Atomic Energy Basic Act, all licensees must establish Operational Safety Programs, which establish provisions through which to foster a culture of safety, wherein safety is the first priority of utilities; this program must subsequently be approved by the NRA (IAEA, 2018). All licensees have primary responsibility for ensuring safety. Similarly, the NRA formulated the "*Statement on Nuclear Safety Culture*" in 2015, which establishes a safety culture that all licensees must follow, and which is based on the following principles: (1) safety is prioritized; (2) all risks are taken into account by decision makers; (3) safety will be fostered, sustained, and strengthened through dialogue with the public and licensees; (4) scientific and technical expertise for safety will be maintained; (5) effective communication will be made in the workplace; and (6) nuclear safety will be harmonized with nuclear security (NRA, 2016, p. 74).

To achieve the "3Es plus S" policy, Japan's NEP has undergone several major changes since the 2011 disaster (Vivoda and Graetz, 2015). These include: (1) the establishment of the Energy and Environment Council in 2011, which provides the information necessary to balance energy efficiency and environmental protection; (2) the NRA, which was established in 2012 under the Ministry of the Environment as a new independent commission to protect the general public and the environment through the rigorous and reliable regulation and oversight of nuclear activities (IAE, 2016); (3) the Japan Nuclear Safety Institute was established in 2012 to increase NPP safety and to mitigate the risk of future accidents; (4) the RRA was amended based on safety standards proposed by the IAEA; (5) the Reactor Safety Examination Committee for investigations and deliberations on the safety of reactors and nuclear accident, and the Radiation Council for deliberations on technical standards for radiation hazard prevention were established in 2012; (6) a Nuclear Emergency Preparedness Commission was established in 2012 which enhances measures to prevent nuclear disasters; (7) with the amendment of the Prevention Act, licensees are obliged to conduct periodic safety assessments to ensure the continuous improvement of nuclear reactors throughout their lifecycles; (8) with the amendment to the RRA, a safety assessment system to ensure the enhancement of nuclear safety was introduced in 2012—all NPPs were required to meet the new regulatory requirements to obtain authorization for their continued operation, which consists of three processes: "Application for the Amendment of Reactor Installation Permit," "Application for Construction Plan Approval," and

"Application for amendment of Operational Safety Programme Approval" (METI, 2015).

All these reforms indicate that safety measures have been strengthened following the FNA, which represents good progress in terms of ensuring nuclear safety and the attempts to avoid further accidents. These reforms, however, do not mean there is *no* future risk of accident or incident in Japan because safety and zero risk are not—as some people believe—synonymous. No matter how advanced nuclear technology is or might be, an element of risk always exists. Risk can be mitigated through safety standards (Roeser, 2011), but not eliminated despite sophisticated nuclear technology (van der Pligt et al., 1982). In order to achieve nuclear safety, six reactors in TEPCO's Fukushima Daiichi NPPs, one reactor in JAPC's Tsuruga NPP, two reactors in Kansai Electric Power Company Mihama NPP, one reactor in Chugoku Electric Power Company Shimane NPP, one reactor in Kyushu Electric Power Company Genkai NPP, and one reactor of Shikoku Electric Power Company Ikata NPP—twelve reactors in total—were permanently shut down after compatibility assessments based on the new safety and operating regulations (NRA, 2016).

However, while the new safety standards introduced by the NRA address design-based standards, severe accident countermeasures and standards designed to mitigate damage associated with earthquakes and tsunamis—regulations require utilities to maintain robust tsunami defenses, including seawalls and watertight doors—there has been relatively little emphasis placed on determining whether and how nuclear facilities might be targets for terrorism or war. Natural events are not the only risk to nuclear safety (Strupczewski, 2003) for two main reasons. Firstly, Japan is one of the safest countries worldwide in terms of nuclear safety (Komiya, 1999), but as a country has suffered its own share of terrorism, such as the Tokyo subway sarin attack in 1995; indeed, Japan is one of the rarer countries which experienced an act of domestic terrorism using chemical materials (Box and McCormack, 2004). Secondly, Japan is the only country against which atomic bombs have been used in times of war. Its nearest neighbor, North Korea, which has nuclear weapons, has previously threatened Japan, and indeed fired missiles over Japan in 2017 many times (Cha and Kang, 2018). This means that Japan's NPPs remain vulnerable to terrorist attack and potential war, which appear to be the two main weaknesses in Japan's new safety regulations that have not, to date, been suitably addressed.

Additionally, although there are international standards on nuclear safety that Japan adopts (Kingston, 2014a), "how safe is safe enough?" may be different for the public than the safety regulations that the NRE ultimately implements (Yamakawa and Yamamoto, 2016). Research on risk perception shows the public may perceive environmental risks slightly or significantly differently from experts, scientists, and officials (Sjöberg, 2000; Bier, 2001). Safety standards may be well organized and implemented, but if the public does not feel safe with NPPs and the associated safety standards, the policy may nevertheless remain ineffective because the associated fear may cause mental health problems (Kingston, 2013a). Indeed, this seems to be the case in the aftermath of the Fukushima accident because Japan's NEP does not address how people's concerns, priorities, or risk perception are reflected in safety regulations. There may be a need to determine the public consensus on nuclear safety in Japan, where the public trust in nuclear energy is already low (Figueroa, 2013).

The Decrease in Dependency on Nuclear Energy

The second important policy change due to the FNA was to decrease dependency on nuclear energy. The Japanese government forecast a significant increase in the share of NPPs to 50% of total electricity supply in its 2010 Basic Energy Plan through the creation of fourteen new nuclear reactors by 2030 (METI, 2015); however, the FNA changed its longer-term energy strategy in a way that NPPs will now be reduced from 30% to 22% of the country's total electricity production in the same timeframe (IAE, 2016). METI (2015) established Japan's new energy policy in a report called the Long-term Energy Supply and Demand Outlook where nuclear energy accounts for 20–22% in terms of the total power generated for FY2030, implying that nuclear will remain one of the important options for Japan's energy policy as it will continue to form a significant share of Japan's total power requirements.

According to Japan's NEP documents, there are two reasons that all NPPs were not shut down in the short term (METI, 2015). First, the country's self-sufficiency in terms of primary energy sources, excluding nuclear energy, is only 6% (JAEC, 2015). Electricity tariffs rose by about 30% between 2011 and 2012, which cost 4 Trillion Yen of national wealth (JAEC, 2015). The goal is to lower electrical power costs for reasons of economic efficiency. This, however, does not mean that NE is not itself costly, as the above consideration does not include environmental costs. Japan's NEP is not tasked with addressing issues related to the costs of environmental remediation and social impacts. Implementing the new standards required for the operation could be extremely costly; the electric power companies expect the cost of implementing the new standards to be in the region of $11.75 billion (Kingston, 2013b). Second, CO_2 emissions rose by a further 6.7% in 2012 when most of Japan's nuclear reactors were being kept offline (Homma and Akimoto, 2013). However, the role of NPPs on CO_2 emissions is somewhat overestimated because the overall share of Japan's greenhouse gas emissions from NPPs is approximately 30%, which means 70% of Japan's greenhouse emissions have nothing to do with nuclear energy, but with transport, heating, and so on (Vivoda, 2012). For this reason, the country's policy is to gradually decrease dependency on nuclear energy between now and the 2030s, while simultaneously achieving the policy goals related to 3Es plus S.

There is, however, considerable uncertainty surrounding the future direction of energy policy and regulation. It is not known whether Japan will construct NPPs after the 2030s. It seems that Japan's current policy is to decrease dependency on NPPs, but this is ultimately due to the fact that some NPPs did not meet new safety standards, as explained above. In other words, this apparent decrease in dependency is the result of new safety regulations. Japan's NEP states that dependency on nuclear energy will be reduced, but it is not known how this will be achieved or indeed whether NPPs will be closed. Thirdly, the manner in which public demand about nuclear energy is reflected in this policy is now clear. Public trust in nuclear energy is not high, but the government is still showing a tendency to continue with NPPs, which may create tension between the government and the public over nuclear-related issues as underlined by various researches (Kingston, 2013a; Vivoda, 2016).

Renewable Energy

Japan prioritized renewable energy in order to achieve the 3Es plus S policy after the FNA. Ministry of Economy, Trade and Industry (METI) announced Japan's new energy policy in its Energy White Paper, as published in 2017. According to this policy document, the renewable energy target for 2030 is 20–24% of total electricity generation. This seems credible because the average annual growth in renewable energy was 5% between 2003 and 2009, and 9% between 2009 and 2012, while it made a significant annual increase of 29% between 2012 and 2015, the highest on record. This means Japan achieved a rapid transition to renewable energy after the FNA.

There are two main reasons for renewable energy being prioritized as an alternative to nuclear energy after the 2011 disaster. Japan's NEP documents underline the fact that Japan has poor fossil fuel reserves, and that importing fossil fuels increases emissions and energy prices. Renewable energy appears to represent an excellent alternative by which to improve Japan's self-sufficiency rate and reduce CO_2 emissions because they are sustainable energy solutions that do not adversely affect the environment (Quaschning, 2009). Renewable energy has, however, some disadvantages compared to NE. Firstly, it is a costly energy source—while 1 kW h costs roughly 5 yen in nuclear energy, it costs 14 yen in the wind and 46 yen in solar energy (Kingston, 2013a; Koyama, 2013). Secondly, renewable energy sources still have issues relating to supply stability, for example, energy cannot be generated on rainy days or when the wind does not blow (Bevan, 2012), which is, of course, not the case for nuclear energy. Japan has four seasons and renewable energy cannot generate electricity at the same levels throughout the entire year (Vivoda, 2016). Due to these disadvantages, Japan's new energy policy aims to diversify energy sources through an optimal combination of various power sources in order to simultaneously achieve safety, supply stability, economic efficiency, and environmental sustainability (METI, 2015).

Transparency

Transparency is another key principle in Japan's new energy policy after the FNA. In particular, the policy-making process and the issue of the "nuclear village" have been criticized by the media, NGOs, and the public over the decades (Kingston, 2013a). In order to solve this issue, the NRA established the Policy on Ensuring Operational Transparency in 2012, which stipulates that transparency is one of the basic policies of the NRA. Article 3 of the NRA's Core Values and Principles states: "*We shall ensure transparency and appropriate information disclosure on regulations, including the decision-making process*" (NRA, 2016). This has also been underlined by various policy documents since the accident such as the Energy White Paper published in 2017.

According to the Policy on Ensuring Operational Transparency, the NRA aims to achieve transparency through the following: (1) all information regarding nuclear energy-related activities is disclosed publicly; (2) all meetings between the NRA's commissioners or staff and nuclear energy companies and the outlines of the meetings' discussions are made publicly available; (3) the NRA shares the names of the NRA attendees and the documents used in the meetings; (4) the NRA Commissioner Meetings and other official meetings are open to the public, and real-time video is uploaded to the NRA website available in Japanese and English; (5) the NRA chairman holds a weekly press conference which is made available as live broadcasts, and recordings of which are uploaded to the NRA website; (6) all research plans, data, and conclusions and associated self-evaluations of the research are disclosed; (7) the NRA obliges licensees to periodically make safety reports and risk assessments available to the public; and (8) the NRA provides information on the results of environmental radiation monitoring, including the sea as a monitoring target, at a frequency of once per month (NRA, 2016).

DISCUSSION AND CONCLUSION

The discussion and conclusions sections consist of three parts. They firstly link the interpretation of the evidence to the broader literature, and then discuss the cross-cutting issue between Japan's HRP and NEP.

Japan's HRP

This research has analyzed Japan's NEP and HRP. We have found four main findings. Firstly, the analysis indicates that Japan does not have any policy on human rights in connection with the environment. There is extensive academic study to suggest that Japan has integrated IHR standards into its constitution, such as the right to life and health, which is consistent with this research (Port, 1991; Iida, 2004; Fujioka, 2003). From an EHRs theory perspective, these rights require a safe environment because people are dependent upon the quality of, and access to, the surrounding environment in this regard (Anderson and Boyle, 1996). However, we identified that the connection between the two is not recognized by Japan's HRP in any way. Some scholars argue that EHRs should be addressed at the constitutional or policy level because it will oblige the state to protect human rights through environmental protection (Anderson and Boyle, 1996). Indeed, it may oblige public institutions to protect the environment, as the state has a legal duty to protect human rights. However, the supposed fact that the environment is essential for the enjoyment of human rights is somewhat vague because some human rights, such as the right to religion, may be enjoyed without needing access to a safe environment. There is a need to clarify which human rights specifically require a safe environment.

Parker (2009) underlines a different point, namely, that there is no need for such recognition at the policy or constitutional level because the connection between the two is often constructed through judicial interpretation. This seems a valid argument because the majority of African countries, for example, guarantee a "right to life" in their constitution without referencing the concept of a safe environment but the connection between the two is interpreted by courts in related lawsuits (Parker, 2009). Similarly, courts in many countries such as India, Pakistan, and Colombia recognize the right to a healthy environment as an inherent element of the right to life, even though their constitutions do not actually reference the former. The Supreme Court in Pakistan, for example, ordered that the dumping of nuclear waste creates environmental pollution that violates Article 9 of the Constitution of Pakistan which states, "*no person shall be deprived of life*" (Eurick, 1999). Environmental protection and human rights are, therefore, viewed as distinct elements by Japan's HRP.

As for substantive EHRs, this research found that the right to the environment is not on the agenda of Japan's HRP. The right to the environment is recognized by more than fifty-six countries through constitutions (Boyd, 2011); there is, however, still no consensus on whether the right to the environment should be added to the constitution as a separate human right. One group state that the right to a safe environment is a genuine and ecologically sound human right that should be provided for in the constitution (Hayward, 2005; Thorme, 1990) because it would place states under obligation to secure a clean environment, as the responsibility to protect human rights resides first and foremost with states themselves (Eacott, 2001; Ziemer, 2001). As a rebuttal to this point, other scholars state that well-established human rights such as the right to life and health inherently require a safe environment, which can be mentioned in constitution/law or can equally be interpreted by the courts (Anderson and Boyle, 1996; Shelton, 2008a). Indeed, as explained above, courts have often ruled that this right is implicit to the right to life, the right to health, and the government's duty to protect the environment. From this perspective, Japan arguably may not need to guarantee a distinct right to the environment at the constitutional level because universal human rights, such as the right to life and health, which require a safe environment are already well established by Constitution of Japan.

In contrast to the right to environment, my analysis indicates that the ICCPRs that Japan has ratified influence the country's human rights framework through the recognition of internationally recognized civil and political rights such as the right to expression, assembly, access to court, and so on, within its constitution. We found that civil and political rights are well established in Constitution of Japan, which is in line with the results of the extant literature (Repeta, 1987; Neary, 2002; Port, 1991). EHRs theory argues that civil and political rights are effective instruments that can be, and are, invoked by citizens to address environmental issues (Anderson and Boyle, 1996; Anton and Shelton, 2011). Japan guarantees a variety of civil and political rights such as the right to protest and to petition that can be used by concerned citizens in environmental matters. This, however, does not necessarily mean that they are recognized or should be recognized specifically as EHRs because these human rights can be used in any public issue related to education, the healthcare system, and so on, and are not purely restricted to environmental matters because the constitution does not currently impose any boundaries on where human rights can be used, and thus people are free to invoke these rights in environmental matters—they can protest about nuclear energy or collect petitions for nuclear energy-related decisions. From this perspective, civil and political rights are HRs explicitly recognized by Constitution of Japan, which is not to say, however, that they are necessarily respected in practice.

As for PRs, the analysis shows that Japan recognizes the right to access to justice and to information, which are defined as PRs by EHRs theory (Anderson and Boyle, 1996). Access to information and justice are two human rights that are guaranteed by various human rights documents, such as the ICESCR, that Japan tends to adopt, and indeed these have already been ratified (Iwasawa, 1998; Beer, 2009). However, neither Constitution of Japan nor ICESCR define these rights as procedural EHRs; only international environmental declarations and conventions such as the Rio Declaration and the Aarhus Convention define them as PHRs (not human rights) that can be used by individuals with regards to environmental matters (Shelton, 2008b). The literature suggests that these human rights are the reformulation and expansion of existing human rights and duties in the context of environmental protection, since the Rio Declaration formulates the link between human rights and environmental protection largely in procedural terms (Parker, 2009; Fitzmaurice, 2002). As is consistent with the literature (Neary, 2002; Iwasawa, 1998; Beer, 2009), we have found that they are recognized as human rights by Constitution of Japan.

Japan's NEP

The second main finding is that the FNA has changed Japan's NEP significantly, which is in line with the related literature (Koyama, 2013; Vivoda and Graetz, 2015; Kingston, 2013a). This research has identified the fact that Japan made radical reforms to its NEP after the FNA. Firstly, safety was prioritized, and new safety regulations were formulated, since the nuclear accident was partly a result of human failure, as found by previous research (Duffield and Woodall, 2011; Vivoda, 2012). Nuclear safety is the top priority of Japan's new NEP, which seems an important reform intended to achieve environmental safety. There has been considerable research into the fact that Japan has rebuilt its safety regulatory system in order to avoid a recurrence of the FNA, including organizational restructuring (such as NRA), the introduction of new nuclear regulation, and enhanced crisis management. This research has simultaneously found that nuclear safety has become the top priority of Japan's new NEP, which represents an important reform intended to increase environmental safety. The difference between my findings and the results of other research (May, 2017) is that terrorist and war-related issues are ignored by the new safety requirements. Japan is one of the safest countries worldwide (Komiya, 1999), but the risk to NPPs associated with terrorism and war will always exist (Strupczewski, 2003). The country's closest land borders are to North Korea and China, who both have nuclear weapons and who are in ideological conflict with Japan. There is a vast body of literature that suggests terrorism and war are the two major threats to

NPPs (Bodansky, 2007). This study found that these issues are not currently addressed by Japan's new safety regulations. However, the question of how a NPP can be kept safe under conditions of war remains unanswered in a far more general sense because military technology is currently highly advanced, and some neighboring countries (again, such as North Korea and China) have the (nuclear) capability to damage NPPs regardless of any measures that might be in place (Murray, 2009). This, however, does not change the fact that Japanese NPPs remain vulnerable to terrorist attack or war, despite the new safety regulations. This vulnerability, therefore, represents an ongoing weakness in Japan's new safety regulations, which has not been reasonably addressed by previous research.

Thirdly, and consistent with previous studies (Chen et al., 2014; Elliott, 2013), this study found Japan has shown a tendency toward decreasing its dependency on nuclear energy while increasing its usage of renewable energy after the FNA. There is a vast body of research studying Japan's shift from nuclear energy to renewable energy since the 2011 disaster (Meltzer, 2011; Homma and Akimoto, 2013). Zhang and Mclellan (2014), for example, stated that the FNA changed the ratio of renewable energy (except hydro) in the electricity supply significantly—while this ratio was less than 2% in 2010, the intention is that it will be 20% by 2030. Similarly, Komiyama and Fujii (2017) noted that the post-Fukushima energy policy places a high priority on expanding renewable energy as the socially preferable option, meaning that citizens tend to see renewable energy as an alternative to nuclear energy. There is, however, still uncertainty about the future of nuclear energy after the 2030s. It is not known whether nuclear energy will continue to play a key role in Japan's energy policy after the 2030s as noted by Vivioda (2016) and Kingston (2013a).

Fourthly, we found that transparency is a new key principle of Japan's new NEP. A number of research efforts have found transparency to be a key reform after the FNA (Elliott, 2013; Vivoda, 2016; Kingston, 2013b) because the NRA disclose all information to, and share meetings with, the public; indeed, the NRA goes to some considerable effort to share all available information with the public (NRA, 2016). Certainly, this seems good progress toward achieving openness and accountability (Kingston, 2013a). However, the policy does not seem entirely consistent with its implementation. There has been considerable research providing evidence that the government's and TEPCO's failure to be transparent, essentially due to their uninformative public release during the accident which noted that the government and TEPCO did not share emergent information with the public at the time of the accident (Watanabe, 2016; Figueroa, 2013). Similarly, Yukiya Amano, director general of IAEA stated that "*sufficient information failed to reach the IAEA in the initial phase of the accident*" (Wang and Chen, 2012, p. 3575). However, nearly eight years have passed since the FNA, and by this stage one might expect that the government and TEPCO have learnt the appropriate lessons from its failure, meaning it might not be fair to label the current governance with previous mistakes or failures, as noted by other researchers (Vivoda and Graetz, 2015; Iguchi and Koga, 2015). For example, all official meetings are open to the public, and real-time video is uploaded to the NRA website and documents regarding decisions and policy are released publicly, which seems to represent good progress toward achieving transparent governance, as underlined by some other researchers (de Saille and Matanle, 2016). This, however, does not mean that transparency is achieved by the current governance. IEL suggests that public institutions should provide information in formats that are easily understood by the public (Gupta, 2010; Ramkumar and Petkova, 2007). In other words, the live broadcast of meetings or release of public reports and decisions are not in themselves sufficient to achieve transparency because lay people may not fully understand the context, decisions, or discussion in the meetings themselves. Nuclear energy is inherently a highly technical topic (Yamamura, 2015), and there may be a need to provide easily understandable information for public consumption. This issue has not been addressed by the NRA in any way, as also noted by other researchers (Kingston, 2014b).

Chapter 3

Violation of Substantive Human Rights in Fukushima

INTRODUCTION

As discussed in chapter 2, a safe environment is essential for the enjoyment of human rights. The relationship between the environment and human rights is particularly relevant in the case of the Fukushima incident because this particular accident released large amounts of the radiation into the environment. Agricultural land, food supply, water resources, the ocean, and the air have been heavily contaminated by the radioactive materials released, but thousands of people are still forced to live in this contaminated environment in THUs. However, how these people's fundamental rights such as the right to life, health, and property have been so affected is still unclear due to a lack of relevant research in this specific area.

In order to fill the above gap in the current knowledge, this chapter presents the environmental impacts and risks of the FNA on/to human rights (particularly the rights to life, health, and property) through TA. The chapter consists of three sections: it firstly analyzes the various stakeholders' approaches and official reports on the impact of the accident on people's lives, health, and property; secondly, it has analyzed Fukushima residents' experiences of the effects that the accident has had on their lives by addressing the first research question, namely, that of what Fukushima residents' experiences were of the consequences, and ongoing environmental effects, of the accident in relation to the right to life, health, and property; lastly, the discussion section links the evidence to the broader literature. The central aim of the chapter is to answer the following question: how does the environmental impact of the accident on people's lives, health, and property relate to human rights?

ENVIRONMENTAL IMPACT OF THE FUKUSHIMA ACCIDENT ON PEOPLE'S LIVES, HEALTH, AND PROPERTY

Before presenting the TA based on responses from residents, this section gives an overview of the environmental impacts of the FNA on people's rights to life, health, and property. Two different data have been used. The first is that of officials and academic reports which have been collected through an online search, while the second was from interviews with officials, TEPCO managers, and human rights and environmental NGOs in Japan.

The Right to Life

According to the basic needs theory, air, water, and food are the most fundamental human needs which must be met to ensure survival. Their complete lack, or poor access to them, may result in death or serious health issues. This means that food, water, and air represent the most important and basic needs for individuals to realize and enjoy the right to life. The FNA, however, has contaminated the entire environment including the air, water, and food supplies which clearly pose an ongoing risk to the enjoyment and realization of the right to life in the region (Eisler, 2013).

According to my data, the risks associated with the FNA have been perceived differently between officials and residents. Firstly, all TEPCO and government officials stated there is no serious risk to people's lives associated with the current radiation levels in 2017 because the current levels around Fukushima are not high, excluding the 20 km exclusion zone. All officials stated the fact that no resident has died due to the FNA to date demonstrates that residents are, in fact, safe. This risk perception is based on the fact that the Japanese government has officially set an annual effective dose limit to the public of 20 mSv for the Fukushima Prefecture and the exclusion zone, and THUs are designed in accordance with the official annual effective dose limit. The majority of the region in Fukushima is lower than the 20 mSv/yr limit. In contrast, all NGOs, activists, and the majority of scholars (ten out of fifteen) that we interviewed disagree that the ongoing contamination threatens people's lives because radiation levels in many areas, such as the village of Iitate, are still too high to live in.

As for food safety, the government monitors and responds to issues regarding radionuclide contamination of food through reference to CODEX-based standards, which is an international standard for radionuclides in foods. All government officials and scholars (three out of three) who work on food security stated that only agricultural products that are confirmed to be safe are sold in the markets, and this information is shared publicly—the informational leaflets which guarantee the safety of Fukushima-produced foods are distributed by the public authorities, and the same information is also available online. Indeed, Japan has a food safety guideline which prohibits the sale of potentially contaminated foodstuff (Higuchi et al., 2017). The risk assessment is, in practice, conducted by the Food Safety Commission, as established under the Food Safety Basic Law. Foods produced in Fukushima are also recognized as being safe by the Food and Agriculture Organization of the United Nations (GOJ, 2018). None of the food which is sold in the markets exceeded the standard level of 100 Bq/kg established by the Japanese government—the Japanese authorities issue appropriate radiation certificates for local food in Fukushima. No issue in relation to food safety has been raised by NGOs or other stakeholders.

Lastly, Japan's current limit for radioactive cesium in water is 10 Bq/kg—for comparison, the U.S. safety limit for cesium levels in drinking water is about 28 Bq/kg. However, there is no current consensus among officials and NGOs regarding water safety in Fukushima. According to TEPCO and government officials, water safety is perfectly adequate, is monitored by the public authorities and appropriate information shared with the public. Indeed, Japan has a Waterworks Law which obliges the public authorities to promote safe water for public consumption shows that there have been no detectable amounts of radioactive cesium or radioactive iodine in tap water to date. However, all environmental NGOs and scholars who work on environmental issues in Fukushima claim that cesium in water has exceeded regulation levels since the time of the accident. This issue has also been accepted by the government, in that the Nuclear Regulatory Agency announced in June 2016 that there was some radioactive contamination from the damaged stations in the tap water of several prefectures, namely, Miyagi Prefecture, Yamagata Prefecture, Fukushima Prefecture (Yontrarak, 2018). This raises the point that the government failed to achieve the required water safety standards—it is not known how many people are impacted, or indeed how long they drank contaminated water for.

The Right to Health

Health assessment regarding the FNA in Fukushima is conducted by the Japanese government in a timely manner. Fukushima Health Management Survey (FHMS) is the largest health monitoring project to have been launched by the Fukushima Prefecture and is implemented by the Fukushima Medical University in order to monitor the long-term health of residents. Official reports, scientific research, and NGOs' reports all indicate that mental disorders are the most common health problem in connection with the radioactive contamination among residents (Kunii et al., 2016). In interviews, all TEPCO and government officials, researchers, NGOs, and activists emphasized the point that mental health issues are very common among residents due, in particular, to the fear of radiation.

According to the FHMS, physical health issues are not anywhere near as common as mental health issues. The reason behind this, as stated by all officials, is that all residents have access to safe water, food, and an environment in Fukushima. According to the FHMS, there has been, however, a slow increase in: (1) obesity after the disaster due to reduced physical activity; (2) the annual incidence of thyroid cancers in the Fukushima Prefecture compared with the those across the rest of Japan; (3) the premature birth rate increased between 2011 and 2012 but dropped in 2013; and (4) cases of liver disorders, hypertension, renal failure, cholesterol or triglyceride, and diabetes in Fukushima (IRSN, 2016). This is consistent with the related academic research that there has been no serious increase in physical health issues in relation to the radiation (Jimba et al., 2013). However, the criticism leveled by scholars and environmental NGOs is that the accident may have a long-term impact on people's physical health as some cancers may begin to develop up to 20–50 years before diagnosis. So, it would seem far too early to assume all the consequences to physical health due to the FNA have already become apparent.

The Right to Property

According to official statistics, 160,000 people were evacuated from their homes, and indeed 50,000 victims currently remained displaced due to government concern about radiological effects from the accident. 1,220,360 buildings in twenty prefectures have been damaged. Forty-two percent of damage to buildings comes from the associated earthquake, 39% from the tsunami, and 19% from the nuclear disaster. All TEPCO and government officials stated that the decontamination process is still underway, and evacuees will be able to return to their homes gradually. Two further steps have been taken by the public authorities with regards to the property issue. Firstly, TEPCO pays compensation for any damage to property as a result of the FNA until decontamination process has been completed. Secondly, the government provides free housing units for those who were evacuated due to the radiation.

These two steps have been criticized by all NGOs and most scholars (eight out of fifteen) for two reasons. Firstly, the amounts of compensation are extremely low. TEPCO recommends the negotiation of compensation on an individual basis, which means each person or family may receive a different amount of compensation to others because payments vary depending on the amount of radiation recorded in a particular area. TEPCO pays 8,000 yen (£54) compensation per night until victims are able to restart their lives in their former hometowns (TEPCO, 2018). Secondly, the THUs are not comfortable enough to live in because they are small, are poor quality, and are isolated from the city centers, as well as being close to the damaged stations.

Free housing units have indeed been provided to all victims of the FNA. Additionally, TEPCO is obliged to pay compensation and, indeed, does. Besides, residents do not pay any rent for the THUs and the residents will return to their homes after the cleaning up process is complete. This, however, does not mean that the free housing units and the compensation meet people's accommodation needs. We visited more than ten THUs, all of which were small in size and are isolated from the town. Residents do not have any immediate access to supermarkets, gyms, ATMs, and so on, around the free housing units. It is clear that they do not have as comfortable accommodation as they used to have prior to the accident, despite the availability of free housing units and compensation.

On the other hand, there is no current academic research regarding property issues in relation to the contamination in the evacuation zone. All scholars and NGOs stated that they have no detailed information about the property situation in the exclusion zone. There are two main reasons: first, there is still no permission to enter into some parts of the exclusion zone due to high radiation levels; and second, it is too early to determine the condition and value of the properties because it is still not known when the decontamination will be complete and there is no information about the condition of the property in the exclusion zone. All these demonstrate that the accident has contaminated the environment, which in turn has affected food and water sources, people's health, living conditions, and property. The following section analyzes Fukushima residents' experiences of the ongoing contamination on their lives, health, and property.

HUMAN RIGHTS ISSUES REGARDING RADIOACTIVE CONTAMINATION IN FUKUSHIMA: TA

The policy analysis in chapter 3 shows that Japan legally guarantees internationally recognized human rights. Recognition within the law in isolation is not, however, sufficient to realize human rights because factors such as the lack of a safe environment may render people unable to realize and enjoy their human rights in practice (Hajjar Leib, 2017). This is particularly the case in the FNA because the 2011 accident heavily affected the environment in which thousands of people still live; this means that the contaminated environment may pose risks to, or affect the realization of, human rights in Fukushima. In order to investigate how the Fukushima accident has affected basic human rights, this section analyzes residents' experiences of the consequences and ongoing environmental effects of the accident in relation to the right to life, health, and property in the current THUs through TA by addressing the first research question. The analysis is based on interviews with thirty-four residents who live in THUs.

TA has generated three overarching themes and eight subthemes. All responses are ordered in the table from higher to lower answers. Percentages refer to the number of people who mentioned each given theme. Nine subthemes are categorized under three themes including "Violation of the Right to Life," "Violation of the Right to Health," and "Violation of the Right to Property" in light of Japan's HRP, for example, Article 16 of Constitution of Japan states: "*We recognize that all peoples of the world have the right to live in peace, free from fear. . .*"; for this reason the subtheme "Living with Fear" is categorized under "Violation of the Right to Life" theme. A detailed description and interpretation of the themes are provided in the subsequent text, along with quotations from the interviews.

Theme 1: Violation of the Right to Life

The "Violation of the Right to Life" theme refers to how the current environment restricts and affects the realization and enjoyment of people's right to life. No resident has died in Fukushima due to the radioactive contamination to date, but most of the residents referenced the issue that the ongoing contamination has posed risk to, and has had direct impacts on, people's lives since the 2011 disaster. All residents (thirty-four out of thirty-four) stated that they are currently in unsafe living conditions because of the contaminated environment in Fukushima, which ultimately restricts their lives. The residents referred to the issue of living with fear, which in turn poses risks to, or has affected, the realization and enjoyment of the right to life as explained below.

Subtheme 1: Living with Fear

All residents stated that they are afraid of living in the contaminated environment, which is near the damaged stations and the 20-km exclusion zone. Thirty-four out of thirty-four residents stated that they believe that they could easily be exposed to the radiation and then suffer from serious physical health issues such as cancer and leukemia if they continue to live in this potentially unsafe environment. This particular risk perception is very common among the residents, which in turn triggers the fear that they might be affected by radiation-related illness or even die due to the ongoing contamination in the region. In particular, parents are much more afraid than the younger generation because they believe that their children are more vulnerable to radiation, even though officials have stated that "*the radiation levels in the region do not result in any greater exposure than the normal dose for the region*" (Official 1, June 27, 2017).

Living with fear poses a serious risk to, or affects the quality of, lives because residents consciously avoid doing many social and daily activities due to the risks associated with the radioactive contamination. They believe that they could come into contact with radiation when they spend large amounts of time outside due to the environment being contaminated by the radiation. For this reason, residents often stated that they avoid having picnics, farming in their gardens, playing games such as football, basketball, volleyball, and so on, in the open areas, or swimming, snowing, hiking, or generally enjoying the green spaces, and so on. This fear of radiation restricts the amount of time that they spend outdoors and engaged in social activities, which is an obstacle to the enjoyment of their lives. Resident 1 (June 5, 2017), for example, stated: "*We are always afraid and depressed due to living near the contaminated environment. For this reason, we avoid going out, eating local goods, drinking tap water, exercising outside, etc.*"

On the other hand, fear is not a risk to only the right to life. Fear also poses risks to health because living with fear triggers mental health issues among residents. However, the issue in relation to fear is that radiation may not be the only source of fear. Japan has a pronounced risk of earthquakes and tsunami and the potential for more natural accident may trigger fear among residents as they faced with one of the worst natural accidents in 2011. This is also underlined by the residents. Fourteen out of thirty-four residents stated that they are afraid that they might have to face tsunami and large earthquakes again. Resident 5 (June 9, 2017), for example, stated: "*Radiation is our main concern. We have a fear that we might be exposed to the radiation. However, we are also afraid that earthquake and tsunami will happen again, which in turn could affect our daily lives.*" This shows that radioactive contamination is the main reason for this fear, but there are clear difficulties to measuring the extent to which the contaminated environment is causing fear among residents.

The issue of living with fear is also underlined by all NGOs and scholars. They stated that residents tend to perceive current radiation levels as a great risk to both their lives and health, which in any case significantly affects the quality of their lives as they avoid doing certain daily activities such as gardening, hiking, walking outside, and so on. NGO2, for example, stated: "*People's main concern is radiation. They live with the fear that radiation will affect their health.*" However, TEPCO and government officials hold the opinion that the risk inherent to the contaminated area has been prevented through evacuation and the current THUs were constructed in a safe location where radiation does not pose any serious risk. It seems that risk perception differs from person to person. While residents, NGOs, and scholars define current radiation levels as risky, officials and TEPCO managers do not regard them as hazardous to human health. This means that there is a lack of socially acceptable risk which in turn creates conflict between government and the public.

Theme 2: Violation of the Right to Health

Nearly all participants made reference to the health issues that they have had since the accident in 2011. The "Violation of Right to Health" theme references how radioactive contamination has affected people's health in the THUs since 2011. The analysis indicates that residents mostly suffer from similar health issues as they share a common environment, one which has, of course, been contaminated by radiation. People's main concern is that the contamination could cause cancer-related illness, but this is not the case in Fukushima now. My data shows that mental health issues are much more common than physical health issues in the THUs. This, however, does not mean that there is no risk or low risk to people's physical health in relation to radioactive contamination. The accident is a relatively recent event, which means physical health issues may emerge in the medium or long term as the contamination may last for hundreds or thousands of years.

On the other hand, particularly mental health issues are directly connected with risks to people's lives (first theme) including living with fear, which has mainly affected people's psychological health. Six subthemes emerged from the overarching theme, which include (1) mental health issues; (2) physical health issues; (3) loneliness; and (4) social isolation, (5) risks associated with water contamination to health, (6) risks associated with food supply contamination to health, which are all the result of the ongoing contamination of the environment in the region.

Subtheme 1: Mental Health Problems

The "Mental health" subtheme refers to the wide range of psychosocial health conditions that affect people's thinking, reactions, behavior, and mood. Thirty-two out of thirty-four residents stated that they have mental health problems due to the radioactive contamination in the region. The most common mental health issues that residents face are stress, depression, anxiety, and panic attacks. The radioactive fallout has brought anxiety, stress, panic attacks, and depression due to the potential physical effects of radiation exposure. People are understandably afraid of the potential risk this poses to their lives, which causes psychological issues. They believe that they are consuming contaminated food and water, and that the entire environment in which they live in can cause radiation-related illnesses such as cancer and leukemia, and which affects their mental health in the short term.

Five out of thirty-two residents stated that they faced psychological problems, including depression and anxiety, due to personal reasons while twenty-seven out of thirty-two residents stated that they had never faced any mental health issues before the accident. All these residents stated that they had not received any compensation or professional assistance from TEPCO or the government to help deal with mental health issues. Ten out of thirty-two residents have visited psychologists, while and twenty-two out of thirty-four residents still go to their psychologists regularly for assistance. The use of medicines to help with mental health issues is very common among residents, as Resident 10 (June 15, 2017) stated: "*I have been using medicine for depression. I have two children and I feel very anxious about their health and future*" twenty out of thirty-two residents stated they have regularly used medicine for mental health issues since the accident in 2011.

This, however, does not necessarily mean that radiation is the only reason behind these mental health problems. Residents have witnessed tsunami and earthquake, which might understandably represent another source of stress, depression, and anxiety. For example, Resident 24 (June 20, 2017) stated: "*After I lost my baby in the temporary housing unit I started suffering mental health issues. I think that radiation killed my unborn baby.*" Accordingly, it seems that radiation is one of the reasons for this resident's mental health issues, but losing her baby is the main reason. Hence, it is difficult to prove to what extent, or even how other factors such as tsunami, the poor living conditions, and so on, have affected people's mental health in Fukushima.

Scholars raise similar concerns that other factors such as tsunami, earthquake, the loss of family members in the triplex accident, and evacuation have also affected people's mental health. Scholar 9, for example, stated: "*Empirical research indicates that mental health problems are common in the THUs. However, it is hard to say that the radioactive contamination is the only reason because poor living conditions and fear of earthquake and tsunami also affect people's mental health.*" In contrast, NGOs stated that Japan is prone to tsunami and earthquake, meaning that Japanese people have lived through similar natural events and this does not appear to have had any serious impact on people's mental health. However, the FNA is the only nuclear accident in Japan and residents' main concern is radioactive contamination which in turn affects their mental health situation. Officials stated that mental health issues are common among residents due to the evacuation and triplex accident, but they provide free health services to ensure their citizens' mental and physical health is protected. All the above show that there is a consensus that mental health problems are common but no agreement over what exactly causes mental health problems.

Subtheme 2: Physical Health Problems

The "Physical health" subtheme refers to an enduring health problem that will not go away. Thirteen out of thirty-four residents stated that they have had physical health problems in relation to radioactive contamination. Physical health problems that residents have faced include diabetes, hypertension, pregnancy issues, obesity, cancers, and thyroid disorders. The types of physical health problems, however, depend on people's ages. For example, while elderly people reported diabetes and hypertension which have been aggravated by stress and depression in relation to the radiation, young mothers have pregnancy issues because being exposed to iodine 131 slows thyroid and hormone function. Resident 3 (June 7, 2017), for example, stated: "*I have had hypertension issues since 2012. My doctor said to me that stress triggered my hypertension. I was very healthy before the nuclear accident.*" This shows that the mental health issues that have resulted from the contamination have also tended to cause physical health issues.

Physical health problems are less common than mental health problems in the THUs for two principal reasons. Firstly, people make considerable effort to avoid being exposed to the radiation, and in such a way that they do not spend too much time outside, or drink tap water; they also have regular free health checkups provided by the government. Secondly, the Chernobyl nuclear accident shows that cancer-related illness is the most common physical health issues to arise from such incidents (Nikiforov, 2006), but this is not the case in Fukushima because many cancers begin to develop 20–50 years before diagnosis (Williams, 2002). The Fukushima accident happened in 2011, so it may be some time before cancer-related illnesses emerge in the region if residents continue to be exposed to the radiation (Kamiya et al., 2015).

On the other hand, the main issue in relation to physical health is to identify what exactly has caused given illnesses. For example, four out of thirty-four residents stated that they suffered from cancer after the accident, but this does not necessarily mean that radiation was the cause. Similarly, Resident 12 (June 10, 2017) stated: "*My wife was pregnant in 2012 but our baby was stillborn. . . . The doctor did not say anything about why we lost our baby, but I believe that my wife was exposed to iodine 131 due to the contaminated environment in which we live.*" There are numerous factors which can affect people's physical health such as genetics, smoking, drinking too much alcohol, stress, and so on. It does not seem possible to measure the extent to which the ongoing contamination contributes to physical health issues.

Scholars, NGOs, and officials are in consensus regarding physical health issues not being as common as mental health problems. However, scholars and NGOs underlined the point that nuclear energy has long-term impacts on people's physical health, meaning that radiation-related physical health issues might begin to appear more in the future. NGO1, for example, stated: "*We have some evidence that physical health issues in connection with the FNA have increased significantly. However, these are not as common as mental health problems. Thyroid cancer, obesity and hypertension are common physical health problems raised to the ongoing radiation. Radiation-related physical health issues might be more common in the future.*" Indeed, this was the case in Chernobyl, as evidenced by recent empirical research indicating that radiation-related cancers have increased in neighboring countries recently due to the Chernobyl nuclear accident in 1988 because cancer cells can form up in 20–30 years after exposure.

Subtheme 3: Social Isolation

The "Social Isolation" theme refers to the idea that residents lost their social ties after the nuclear accident. Social isolation was a very common issue raised by residents in the THUs. While residents had close neighbors, friends, and colleagues in their previous accommodation all moved to different THUs or cities after the evacuation. Resident 9 (June 8, 2017), for example, stated: "*I had many good neighbours in my previous home but I have not seen them since the nuclear accident. My family feels that we are isolated here.*" According to the residents, they lost their social ties due to radioactive contamination, which has resulted in social isolation. This is another common public health issue in the temporary housing caused by the accident.

Twenty-nine out of thirty-four residents stated that they do not frequently see their former neighbors, friends, or colleagues anymore as they now live in different THUs, which is the result of the radioactive contamination in their previous towns. In other words, the radiation in the exclusion zone forced them to move to different places. However, this does not mean that THUs are isolated or residents do not have the opportunity to build new social ties within them. THUs have common areas where residents can get together and meet new people. All such houses have been constructed in close proximity to each other, which in itself may enable people to form closer relationships with their neighbors. For example, five out of thirty-four residents stated that they lost their previous social ties after the accident but made new social ties in the THUs. For this reason, they do not feel socially isolated anymore. Accordingly, it can be said that the radiation has triggered social isolation, but that it is not an ongoing obstacle to building new social environments that include neighbors, friends, and colleagues. This is also underlined by Resident 31 (August 25, 2017), who stated: "*We lost our social environment but THUs are great places to build new social ties as we share similar issues and concerns.*"

In contrast to residents, TEPCO managers and officials stated that the THUs were designed to build new social ties. The THUs have common places where people can do common activities and socialize. Additionally, there is no obstacle to them contacting previous neighbors, relatives, friends, and colleagues. By contrast, scholars said that the fear of radiation triggers social isolation because people do not want to spend any significant time outside. Indeed, the THUs have common areas that people can get together and spend time; however, this does not mean that they can build social ties again, because people do not want to spend time outside due to the fear of radiation, as stated above. When we visited the THUs, we witnessed the fact that common areas are not used by residents because they now prefer to stay at home due to the fear of radiation.

Subtheme 4: Loneliness

Loneliness is common among elderly people in the THUs. Of the ten interviewed participants aged between sixty and ninety-five, they all stated that they feel lonely in the THUs. All of them stated that they used to live in the same apartment, or at least close to their children in the same neighborhood, before the accident; in other words, they did not use to feel lonely before the accident. They stated that they started to feel lonely after the FNA. The ongoing contamination changed the situation in a way that their children and relatives moved to different cities after the accident. This issue is the result of radioactive contamination because their families and relatives do not want to visit them in their current housing due to the fear of radiation, which is an indirect impact of radiation on people's health. Older people commonly live alone and feel lonely, which is a common public health issue in the THUs. Resident 11 (June 12, 2017), for example, stated: "*I have two children, but they do not want to visit me here because they are afraid that they can be exposure to the radiation. Unfortunately, I live alone here.*"

Another reason behind the associated loneliness is that older people previously enjoyed good social ties in their former neighborhoods, but do not have them in the THUs, as explained above. Ten out of thirty-four residents stated that they used to have close relationships with their neighbors in their previous accommodations, but they lost their connections with them after the FNA, which has contributed to their feelings of loneliness. Accordingly, social isolation has triggered loneliness. However, four out of thirty-four residents stated that their relatives regularly visited them in the THUs. This means that radioactive contamination only discourages people from visiting their relatives or friends in the THUs, but is not an actual obstacle to doing so.

Loneliness is, however, a common issue, particularly among elderly people, all over Japan. According to empirical research in this area, the prevalence of loneliness among older adults ranges from 10% to 29% in Japan (Takagi and Saito, 2015, p. 354). It can be argued that radioactive contamination is only one of the reasons that has led people to feel lonely. In other words, there may be a number, or possibly many other reasons of why people feel lonely. It is difficult to measure the extent to which radiation has caused loneliness in the THUs.

Subtheme 5: Risks to Health Associated with Water Contamination

Water contamination is another major concern that residents referenced. Thirty out of thirty-four residents do not drink tap water because radioactivity has been washed into the water supply such as rivers, lakes, and groundwater. All residents stated that they were worried that they did not have access to a secure source of water. Thirty out of thirty-four residents stated that central government officials and TEPCO managers have informed them that there is no risk associated with the water supplies, while some NGOs and antinuclear organizations informed them that water supply in Fukushima contains radioactive contamination, which of course could be dangerous to human health. Residents tend to believe NGOs rather than the government, ultimately resulting in a lack of consensus over water safety between the residents and government because "*NGOs, scholars and government officials give different information related to the risks associated with the water to people in the temporary housing units*" (Resident 33, August 26, 2017).

Although residents mostly drink bottled water in Fukushima in order to mitigate the risks associated with contamination, the tap water is, however, commonly used for other purposes in people's daily lives such for cleaning, taking a shower or bath, and so on, which is clearly of concern to residents. Thirty out of thirty-four stated that they used tap water because they were unable to afford enough bottled water for cleaning purposes. Resident 19 (July 18, 2017), for instance, stated: "*I buy water from supermarket to drink, but I use contaminated tap water for cleaning purposes as we do not have the opportunity to have safe water here.*" Four out of thirty-four residents stated that although they believe that the water they are forced to continue to use it for drinking and cleaning purposes is not safe, further stating that they cannot afford bottled water for cleaning purposes. However, none of the residents noted any physical health issues in relation to the use of tap water for either drinking or cleaning purposes. Additionally, all government officials and TEPCO managers stated that water safety is being measured by experts, and that the radiation levels in the water fall within the requirements of international standards.

Subtheme 6: Risks to Health Associated with Food Supply Contamination

One of the main issues that residents raised is the lack of secure access to safe food (i.e., no or low amounts of radioactive contamination). Residents do not consider the local food to be safe due to the spread of radioactive material through the food chain. They believe that the accident has released radiation into the atmosphere and radioactive material falling from the air or carried in the rain, water, or snow can deposit on the surface of foods like fruits and vegetables. Additionally, radioactive water has continued to flow into the Pacific Ocean where fish and seafood could take up the radionuclides—Japan is one of the largest seafood consumers worldwide. For various reasons, the majority of residents find consuming local food to represent a risk to their lives and stated that they consume less food to mitigate the risks associated with its contamination. Resident 13 (July 12, 2017), for example, stated: "*I believe that what we eat is not safe here. I was 65 kilograms before the accident and I am 48 kilograms now because I try to eat less to protect myself from radiation.*"

However, residents have noted different types of food which could be contaminated. Five out of thirty residents stated that they avoid consuming seafood as radioactive contamination has been released into the ocean since the time of the accident. Seven out of thirty-four residents stated that they avoid consuming local agricultural products alone, such as rice and mushrooms, and so on, because they believe that radiation has contaminated agricultural land around the damaged stations. Eighteen out of thirty-four residents stated that they do not consume either local agricultural or seafood. Four out of thirty-four residents stated that they do not consider the local food safe to consume due to the radiation, but are forced to do so as they cannot afford to buy imported food, which is, of course, more expensive than local produce. This shows that people perceive food safety in a subjective manner, although the majority of them believe that none of the local food is safe to consume. This, however, does not mean that sea or agricultural food is contaminated by radiation. As stated above, the country has food safety guidelines which prohibit the sale of potentially contaminated foodstuffs, but how safe is safe is a subjective judgment that depends purely on people's values. Official 1 (June 27, 2017), for example, stated: "*we measure radiation level on all food supply and our citizens consume only safe food.*" For this reason, it can be said that public risk perception of food differs from officials' risk standards. This is also underlined by residents. Resident 27 (August 5, 2017), for example, stated: "*We do not agree with the Japanese government, in that the local food supply is not safe due to radiation.*" This represents firsthand evidence that risk perception is subjective. Even if the food is safe or uncontaminated, the fact that residents believe that they are eating contaminated food may, in any case, affect their mental health, which is very common among residents

Theme 3: Violation of Right to Property

The "Violation of Right to Property" theme refers to how the contaminated environment has affected, and posed risks to, people's right to property in Fukushima. Twenty-five out of thirty-four residents stated that they have their own property in the exclusion zone. Nine out of thirty-four participants did not have any property in the exclusion zone—they rented houses but had not bought their own properties. All owners stated that the nuclear accident and contaminated environment has affected their use of their properties significantly. One theme emerged from the overarching theme. This included: (1) "No access to property."

No Access to Property

The "No access to property" theme refers to the fact that residents do not have permission to use their properties due to the high levels of radioactive contamination within the exclusion zone. All residents in the THUs stated that they have no access to their properties such as homes, land, workplaces, and so on, in the exclusion zone since the evacuation in 2011. The authorities created a 20-km exclusion zone around the damaged station, effectively displacing 160,000 people (Nishino et al., 2011). Resident 7 (June 7, 2017) stated: "*My family was forced to evacuate our house in 2011 but I have not seen my house since the accident happened. It was said to us that the area is not safe to live in.*" This means that people who lived in the exclusion zone were forced to leave their property soon after the accident. The area is being decontaminated, though people still do not have access to their property to avoid any risks associated with the radiation.

Ten out of thirty-four residents stated that they only had workplaces in the exclusion zone including grocery shops, barbers, gyms, and so on, which they ran themselves. In other words, they lost both access to their property and their means of making a living. Twenty out of thirty-four residents stated that they only have houses in the exclusion zone. Seven out of twenty residents have houses with their own gardens and garages, while the remainder has only flats in shared apartments. House feature tend to vary by family, while five of them have duplex houses. Two out of seven residents have triplex houses. Flats in shared apartments have 1–3 rooms depending on the flat. All residents stated that they were unaware of the current value of their property due to the fact that the radioactive contamination will almost certainly have decreased this significantly. This is currently, and will almost certainly remain uncertain, as there is no real information about when decontamination will finish.

Residents are concerned that they may not be able to sell their property in the future due to loss of value which can be an obstacle to obtaining a new property since most of them did not have any assets except their properties which are in the exclusion zone. This also affects people's mental health because residents are continuously stressed and concerned that they have spent their life savings buying a house before the accident for its value to have massively decreased as a result. They currently live in THUs that they find uncomfortable because they are generally small in size and are located away from city centers and close to the Fukushima NPP. Resident 1 (June 5, 2017): for example, stated: "*I do not want to live here but I do not have any alternative because I cannot afford a new home. I wanted to sell my own property in the exclusion zone, but this is not possible due to the ongoing decontamination process.*" For this reason, residents want to buy new homes away from Fukushima, but they do not have the savings for this and cannot sell or rent their properties to buy new ones. Most importantly, residents do not know when they will return to their homes and use their properties again.

All the above indicates that people do not have access to their property due to the radioactive contamination. It is a fact that any man-made or natural accidents can cause displacement or evacuation. Risks associated with the environment to people's properties will always exist and cannot be eliminated. In other words, there is no such instance as "zero risk"; however, risks can be significantly reduced through appropriate safety measures. The key question is actually one of whether TEPCO and the government took such measures to prevent the accident, but ultimately the answer is no because the court ruled that both bodies failed to take reasonable measures to mitigate the effects of the tsunami.

DISCUSSION AND CONCLUSION

The vast majority of research has investigated the impact of the FNA on the environment and public health, as the accident resulted in extensive radioactive contamination (Mizuno and Kubo, 2013) that has impacted human health (Yamashita, 2014; Calabrese, 2011; Yoshida and Kanda, 2012). No previous research, however, has examined human rights in connection with the environment in Fukushima. By taking a human rights approach to the accident, this research provides new evidence that residents' right to life, health, and property have been affected by the contaminated environment. By systematically analyzing qualitative data, we have identified various needs and issues in relation to the human rights under three particular themes, which include

"Violation of the Right to Life," "Violation of the Right to Health," and "Violation of the Right to Property."

Firstly, similarities between these research findings and the literature regarding the health impacts of the FNA suggest that the most common issue in connection with the contamination is that of mental health-related problems resulting from the radiation (Yoshida et al., 2016; Goto et al., 2017; Yamashita, 2014). The TA above indicates that nearly all residents suffer, in particular, from depression, stress, panic attacks, and so on, due to the radiation, a fact which is also emphasized in other literature (Yoshida and Kanda, 2012) and official reports (NRA, 2016; IRSN, 2016). There is, however, still no real certainty about the extent to which the contaminated environment has affected people's mental health. A few empirical research efforts have indicated that evacuation, long-term displacement, and poor living conditions in the FNA are other factors affecting people's mental health (Hasegawa et al., 2015). These issues are also underlined by residents, in that the evacuation process and uncomfortable lifestyle in the THUs trigger depression and anxiety. However, it raises the question of why people were evacuated, displaced, and indeed still live in the THU. Radioactive contamination is the main reason for all these issues, as indicated by the extensive literature in this regard (Elliot, 2013; Suzuki and Kaneko, 2013). From this perspective, the contaminated environment appears to be a main reason behind people's mental health issues. In contrast to this justification, some studies have indicated that the 2011 earthquake and tsunami, which ultimately caused 20,000 deaths, has also affected people's mental health (Yabe et al., 2014). Accordingly, the ongoing radiation may not be the only reason behind mental health problems, as residents witnessed the triple disaster of the earthquake, tsunami, and nuclear accident which will have had a psychological impact, according to previous studies (Kukihara et al., 2014). For this reason, it seems difficult to identify to what extent radioactive contamination, in isolation, has affected people's mental health in Fukushima.

What is more certain is that the reason behind these mental health problems is the combination of these three events. However, almost six years have passed since the earthquake and tsunami (Elliott, 2013), but people still live in fear of radiation from the disaster (Yoshida and Kanda, 2012). Accordingly, it seems likely that radioactive contamination is the dominant reason behind mental health problems, while other factors such as decontamination, poor living conditions in the THU, tsunami and earthquake, and so on, also affect it (Yabe et al., 2014). This is also consistent with the interviews, where all residents linked the radiation and subsequent mental health issues although the official report (NRA, 2016; IRSN, 2016) and some academic research (Monzen et al., 2011; Ishikawa, 2017) have indicated that there is no serious risk to people in the region associated with the environment.

Certainly, the region is contaminated as radiation levels vary between 1 and 20 mSv (Takahashi, 2014) and this will, certainly, affect people's mental health (Goto et al., 2017), but residents' perceptions of risks associated with the contamination currently also has an impact in terms of the effect on their mental health—for example, while officials have declared tap water to be safe for consumption (NRA, 2016), residents believe that it is contaminated regardless, which causes anxiety and stress among the victims of the accident. The main reason behind this is that officials and public risk perception of the FNA differ at different levels. Officials determine safety levels in accordance with national standards but objective risks associated with the FNA have been perceived differently in a way that it is not sufficiently safe. This shows that there is no agreement over which risk is socially acceptable, which in turn affects people's approach to the safety measures in a negative manner. The perception of the risks associated with the contaminated environment can, therefore, be said to be the main reason behind the prevalent mental health problems.

We also found that some residents have physical health issues, but these are far less common than mental issues. A number of residents stated that they suffer from cancer and diabetes in connection with the radiation. This, however, does not necessarily mean that radioactive contamination has caused, or is the only reason behind, these issues. There is no consensus in the literature about physical health problems in Fukushima. One group states the radiation will not have any noticeable impact on cancer-related illnesses in Fukushima (Watanobe et al., 2014). This is, however, open to the debate as all these research efforts have used data provided by the government's FHMS, which is not considered reliable by a number of researchers (Brumfiel, 2013), most notably as some of its contributors receive funding for academic projects on the FNA from the government (Rieu, 2013). In contrast to this, some research indicates that there has been an increase in physical health issues due to the radioactive contamination in the region (Ito et al., 2017). A third group is more generally skeptical, in the sense that it is not certain what, exactly, has caused these problems (Hasegawa et al., 2015; Suzuki et al., 2015). This latter view is consistent with my research, namely, that there is no mounting scientific evidence regarding the environmental threats to physical health. Although residents link their physical health issues and the contaminated environment, they do not have any direct evidence, such as doctors' reports, to support their claims, because ultimately it is not possible to identify the exact causes of cancer-related health problems in any given individual. Stress, consumption of unhealthy food, smoking, poor sleeping habits, genetics, and alcohol consumption may trigger cancer or other health issues (Williams, 2002). Uncertainty about the reason behind physical health issues is, therefore, quite considerable.

Japanese people have a human right to health, as guaranteed by Constitution of Japan and the ICESCR agreement that Japan itself has ratified. The right to health requires that people have the right to the highest attainable standards of physical and mental health (Donnelly, 2002). Under the ICESCR, the right to health is subject to "progressive realization," meaning that the Japanese government must take steps toward achieving this right to the maximum of its available resources, which includes safe drinking water and sanitation, food, adequate housing, and a clean environment. The constitutional provisions, however, remain vague in this regard because they do not define what safe water and food, or a clean environment, actually are, which can be perceived differently—while residents do not consider the food, water, and environment safe, which causes mental health issues, officials take the opposing view. From an EHRs perspective, the above analysis indicates that the perception about the impact of radiations is the main factor which affects the enjoyment of the right to health, because people report mental health problems due to the fear of radiation, although this is not the only reason. Radioactive contamination seems the main obstacle to the enjoyment of the right to health, but the question of how this will, or even could, be proven remains unanswered. As Glazebrook (2009) and Shelton (2008a) note, there are certain difficulties with the issue of causation and proof in environmental matters.

Secondly, this research has found that people live with fear due to the contaminated environment. This finding indicates that risk perception, or the lack of socially acceptable risk, affects people's living conditions in the region, meaning that solely determining objective risks to take measures is not sufficient to protect people's health and lives. The official reports on the FNA indicate that there is no serious risk to, or direct impact upon, people's lives (NRA, 2016), as nobody is yet known to have died as a direct result of radiation, and nobody lives in the exclusion zone. The literature is, however, divided into three groups in this regard. Some scholars argue that there is not a serious risk to people's lives as they were evacuated from the contaminated area where the radiation dose is more than the 20 mSv/yr (Monzen et al., 2011; Ishikawa, 2017). In contrast, another group of scholars has argued that people who live near the FPS are at risk due to exposure to radiation, which can trigger cancer at a level between 5 and 20 mSv and may adversely affect human health and result in deaths (Murakami et al., 2017; Davis and Hayes-Conroy, 2018). A third group of scholars states that it is early to identify the impact of the accident on people's lives because long-term impacts can be devastating even after 10–20 years (Elliot, 2013; Kingston, 2012). The third group seems to have a valid argument, in the sense that some cancers may begin to develop as much as 20–50 years before diagnosis (Einstein, 2007), as evidenced by the Chernobyl disaster which, similar to the Fukushima accident, is classified as a Level 7 event by INES. The risk to life is, however, not limited to purely that of death; unlike official reports, which state that there is no serious risk to people's lives (NRA, 2016), this research provides evidence that the accident has affected the quality of people's lives, because people live with the fear of being exposed to radioactive contamination, which in turn restricts the manner in which they can live their lives, a fact which has been emphasized by the extended literature on the subject (Rehdanz et al., 2015). This study found evidence that fear affects the quality and enjoyment of life in such a way that residents avoid various, simple social activities such as having barbeques, hiking, gardening, and so on, due to their fear of radiation. This means that residents' lives are accordingly restricted in Fukushima due to the contamination.

The right to life is a human right guaranteed by both the Constitution of Japan and the ICESCR. The right to life has two distinct meanings: firstly, it is an absolute right that no person, or indeed the state, can attempt to end one's life (Donnelly, 2002). This is not the case in the Fukushima accident, because the accident happened due to a combination of man-made mistakes and natural events. Secondly, the right to life indicates that the state must take appropriate measures to safeguard individuals' lives through the legislature and by taking steps to protect an individual if his/her life is in danger or at risk (Donnelly, 2002). This definition is particularly relevant to the Fukushima matter, because residents believe that they live with the risk associated with the contaminated environment. Article 13 of the Constitution of Japan gives a more detailed explanation of the right to life, in that: "*We recognize that all peoples of the world have the right to live in peace, free from fear and want*" (KANTEI, 2018). This principle implies that the right to life does not only mean breathing or the mere fact of being alive but also living in peace without fear and want. The constitution, however, does not have any definition of fear—the public may perceive the term fear in an entirely different manner from the intentions of the constitution, or indeed those of officials. Theoretically speaking, a person may feel this same fear even in a known, safe environment that is free of any contamination (Cullet, 1995). According to official reports, the radiation levels are not sufficiently high to present a risk to human life (NRA, 2016), but residents perceive the risk differently; they clearly feel that their environment is not safe, or fit, to live in, which in turn triggers fear. The analysis shows that people live with the fear of the continuous impacts of radiation. At the same time the 2011 tsunami and earthquake event, or other potential accident risks, may trigger fear among residents as has been discussed in previous research (Yabe et al., 2014). For this reason, it appears somewhat challenging to be able to identify the extent to which radiation itself causes fear among residents, as there may be other contributing factors. However, what is certain is that risk perception or the lack of socially acceptable risk is the main obstacle to the enjoyment of the right to life guaranteed by Constitution of Japan.

This research further found that the third-most affected human right is the right to property. Limited research has investigated the property issue in the contaminated area, but which mostly deals with the issue of compensation for damage to property (Yamane et al., 2013; Munro, 2018; Tanaka and Managi, 2016). There is also some research investigating the idea that property values have decreased due to the radioactive contamination in the region (Kawaguchi and Yukutake, 2017; Bauer at al., 2017). Various research has reported that thousands of people still do not have access to their property due to the radiation (Elliot, 2013; Suzuki and Kaneko, 2013), which is consistent with my findings. To the best of my knowledge, this research provides new information to the extended literature regarding the FNA which has not been previously discussed (Takahashi, 2016; Eisler, 2013). Firstly, we found that the lack of access to the property is a barrier to being able to move to a new residence, as residents are currently unable to sell any properties in the exclusion zone. The THUs are provided for low-income residents (Elliot, 2013), who had most of their savings invested in properties in the exclusion zone. As they cannot sell their property in the exclusion zone, they cannot move to the new places of residence. This means that being denied access to their property results in residents having to live in the THU. This is turn leads residents to experiencing fear, depression, and anxiety due to the radiation in that region. That is, the lack of access to the property is not just a mere property issue. The analysis also indicates that the decrease in property values also affects people's mental health. Mental health issues in connection with the radiation have been studied previously (Yoshida et al., 2016; Goto et al., 2017; Yamashita, 2014), as discussed above, but my findings indicate that residents feel anxious and worried about the decrease in property prices in the region as well. This means that property issues in connection with the contamination are another reason behind their mental health issues. Indeed, further empirical research is consistent with their concerns that property values have decreased in Fukushima (Kawaguchi and Yukutake, 2017).

The Japanese people have a right to property, as guaranteed by the Constitution of Japan. Article 29 of the Constitution of Japan states: "*The right to own or to hold property is inviolable*" (KANTEI, 2018). The right to property has two meanings: firstly, it implies that people have the right to use and enjoy an asset; the second is that that people are free to own or transfer their property to third parties without state interference (Pascan, 2012). The analysis indicates that residents do not use or enjoy their properties and are unable to determine how to use them in any way, such as whether to sell or rent them, which is in line with other literature (Elliot, 2013), although there is no legal restriction on selling or buying property in Fukushima. However, the scope of the right to property is not limited. States can put restrictions on the right to property by law, particularly when the right is dangerous to the general interest, such as the construction or reparation of roads or other public or communal works (Allen, 2015; Babu, 2012). It seems reasonable to force or oblige residents to leave their homes in order to avoid the effects of natural or human-made disasters (Elliot, 2013), because the state has a positive duty to protect its citizens from the dangers/risks that annual exposure doses might exceed 20 mSv, which is not safe to public health (Yamane et al., 2013). There is, however, still debate in the literature. Firstly, the lack of clarity within the Japanese human rights framework exists in the sense which the right to property is violated in connection with the natural and man-made accidents in order to protect public health (KANTEI, 2018). Secondly, the fact that people were forced to leave their properties for the public safety does not change the fact that people do not have access to the property due to radioactive contamination that TEPCO and the government caused. It is legally recognized, via the Japanese court itself, that both TEPCO and the government were guilty of negligence in contributing to the extent of the Fukushima nuclear disaster, and has ordered them to pay damages (Acton and Hibbs, 2012); indeed, there is a common agreement in the literature and the official reports released by TEPCO and the government in the sense that they did not take sufficient precautions to prevent the accident (Elliot, 2013; Kingston, 2012; NRA, 2016).

From an EHRs perspective, the analysis and the extensive literature available (Elliot, 2013), and indeed official reports (NRA, 2016), indicate that the contaminated environment appears to be an obstacle to the enjoyment of the right to property in Fukushima. The unsafe environment, which is purely due to the radiation levels exceeding 20 mSv, is the reason for the lack of access to the property in the exclusion zone, and indeed the radiation levels remain relatively high. EHRs theory argues that the environment is a means to the end of fulfilling human rights' standards, such as the right to property (Anderson and Boyle, 1996; Shelton, 2008b). In other words, the quality of the environment affects the enjoyment of human rights. The fact that residents are not able to realize or enjoy their right to property due to the environmental contamination in Fukushima shows that the enjoyment of right to property and the quality of the environment are linked in the sense of the Fukushima accident; this linkage is, however, not recognized by Japan's human rights framework in any way. For this reason, the lack of recognition of the relationship between the quality of the environment and the realization of human rights remains a particular weakness in Japan's HRP.

Human rights laws impose obligations on state parties to respect, protect, and fulfill human rights, which require a safe environment as explained above (Wisadha and Widyaningsih, 2018). For example, the ECHRs is a legally binding human rights document which guarantees ICESCRs and CPHRs, though it does not reference the right to the environment in any way (Antonopoulos, 2018). However, the European court has ruled that enjoyment of Article 8, which guarantees respect for private and family life, has been repeatedly violated by environmental contamination (Fitzmaurice, 2011). The court has instructed state parties to take positive steps to legislate or carry out other preventative action to protect private and family life from environmental contamination because the court interpreted Article 8 as requiring specific positive duties from states, namely, that environmental protection may be enjoyed by citizens (Krstić and Čučković, 2015; Jesenko, 2018). In the case of the *Oneryildiz v. Turkey*, the ECHR indicated that the government violated the right to life under Article 1 of the ECHR because Turkey did not share information regarding the risks and impacts of a methane explosion in 1993 at a municipal rubbish tip in the slum quarter of Istanbul (Caglayan, 2015). Similarly, in the case of *EHP v. Canada*, a group of Canadian citizens claimed that nearby radioactive waste dumping was posing a threat to the enjoyment of Article 6—the right to life—of present and future generations, as guaranteed by the ICCPR. The UN Human Rights Committee recognized environmental damage to be a valid cause of action in this case, but declared the case inadmissible due to a failure to exhaust local remedies (Shelton, 2002; Lytton, 2000). Therefore, environmental protection can become a legitimate objective of states, because states have a positive obligation to take all appropriate steps to safeguard human rights which require a safe environment (Boyle, 2007).

To sum up, this research has found clear evidence that the contaminated environment has been a barrier to the enjoyment of human rights in Fukushima, which has resulted in the violation of human rights in connection with the environment. Previous studies of the FNA have mainly focused on environmental contamination and health issues in relation to the radiation. In this chapter, we took an alternative approach to the accident and focused on the impacts of the accident on, and risks associated with the radiation to, human rights including the right to life, health, and property. The analysis indicates that radiation is the main reason behind human rights violations but is not the only one. There are other factors which affect people's right to life, health, and property that Constitution of Japan otherwise guarantees. This means that the accident is not only an environmental or public health issue, which is how the current research regarding the FNA has approached the accident, but also represents a human rights violation. However, the main reason behind these human rights violations is related to the risk perception. Public authorities apply solely objective risks to determine appropriate measures. How people perceive risks, or subjective risks, is ignored by the government which in turn constitutes a violation of the right to life and health. Determining socially acceptable risk through communication with the public could reduce the impact of the accident with regard to people's rights to life and health. These findings can increase public and institutional awareness about human rights in relation to the environment in Japan as well as encouraging public institutions to guarantee the recognition of the link between human rights and the environment at the constitutional, legal, and policy levels.

Chapter 4

Fukushima Residents' Experiences of Procedural Rights in Relation to Japan's Energy Policy

INTRODUCTION

The Aarhus Convention is the first binding multilateral environmental agreement to define and foster PHRs in environmental matters. PRs which are specified in the Aarhus Convention are based on three pillars: the right to information, the right to participation, and the right to access to justice. Many international environmental documents, including the 1992 Rio Declaration, the 1982 World Charter for Nature, Agenda 21 recognize PRs and inspire state parties to adapt them into their national law to ensure effective environmental governance. These three rights are also recognized as human rights by various IHR documents such as the International Covenant on Civil and Political Rights, the African Charter on Human Rights, and the American Convention on Human Rights.

Many scholars agree that PRs can be seen as an effective way of securing environmental protection and democratizing environmental governance, at least in theory, although their implementation in practice can be suboptimal. The literature indicates that PRs may enable individuals to contribute to the environmental decisions which can reflect different groups' concerns, assumptions, values, priorities, and so on, via public participation in decision-making processes. Furthermore, PRs can be used to inform and educate people about environmental issues and safety, and to help them follow the rationale behind environmental decisions and projects via access to information. Lastly, PRs can be used by individuals to seek justice concerning environmental decisions/projects and their impacts on people's lives via access to the relevant courts.

PRs are recognized by Japan's energy policy, the Constitution of Japan, and the IHR documents that Japan has ratified as discussed in chapter 3. Articles 21 and 32 of the Constitution of Japan guarantee access to information and access to the courts, respectively, as human rights, although this is without specifically referencing the environment. Similarly, Articles 19, 25, and 14 of the ICCPR, as ratified by Japan, recognize the right to information, the right to public participation in decision-making processes, and the right to justice. Japan's NEP documents also recognize access to information related to the environment and nuclear energy. Disclosure of the scientifically proven and easily understandable information is one of the key policies in Japan's NEP. Public participation was one of the key principles of Japan's NEP until the antinuclear Democratic Party lost the 2012 Japanese election soon after the 2011 nuclear disaster. Japan has held public hearings, a referendum, surveys, round-table conferences, and so on, from the first construction of NPPs in 1962 until the 2012 election. Lastly, access to justice is not specifically addressed by Japan's NEP.

There is no comprehensive study to date that has investigated the Fukushima residents' experiences of PHRs (including the right to information, the right to participation, and the right to access to justice) in the Fukushima matter. Accordingly, this chapter analyzes the Fukushima residents' experiences of PHRs in relation to the Fukushima event by answering the following research questions: (1) What are the experiences and attitudes of Japanese citizens toward using their PHRs in the aftermath of the FNA?; (2) What are the opportunities and challenges to the realization of the PHRs in the Fukushima matter that have been raised by the residents of Fukushima?

By critically analyzing different stakeholders' approaches (including officials, NGOs, and scholars) to PHRs, we will gain an understanding of Fukushima residents' experiences of PHRs in the Fukushima matter.

FUKUSHIMA RESIDENTS' EXPERIENCES OF PRS IN RELATION TO JAPAN'S ENERGY POLICY

TA has generated three overarching themes and thirteen subthemes. All responses are ordered in the table from higher to lower answers. Percentages refer to the number of people who mentioned each given theme. All responses are categorized according to the three pillars of PHRs, which include: (1) public participation in the decision-making process, (2) access to information, and (3) access to justice. A detailed description and interpretation of the themes are provided in the subsequent text, along with quotations from the interviews.

Theme 1: Access to Information

The access to information theme indicates whether residents have, or have not had access to information regarding nuclear energy, the FNA, and environmental contamination in Fukushima both before and after the 2011 accident. Thirty-four out of thirty-four residents stated that they have sought and received information related to nuclear energy in Japan after the nuclear accident while only two out of thirty-four residents stated that they had sought and received information related to the nuclear energy before the 2011 disaster. This is the most frequently used PR by the victims of the accident. Eight subthemes emerged from the main theme. While the first three subthemes are related to the quantity of access to information, the other three subthemes are related to the quality of access to information, which are as follows:

Subtheme 1: No Attempt to Access to the Information before the Accident

This subtheme refers to the fact that residents did not make any attempt or seek to access to information related to nuclear energy and the Fukushima Power Stations before the 2011 disaster. We asked: (1) Have you ever attempted to receive information related to any aspect of nuclear energy before the 2011 disaster? and, Did any institutions share information related to nuclear energy with you before the accident? And (2) If so, what kind of information, and how did you receive it? Thirty-two out of thirty-four residents stated that they had never received any information regarding nuclear energy before the 2011 disaster because they did not have any interest in nuclear energy and the FPS before the 2011 accident. They also stated that neither TEPCO nor any public institutions had shared information with the residents before the accident. Resident 4 (June 6, 2017) explained,: "*We did not receive any information before the accident in 2011 because we did not demand any information.*" Only two residents stated that they had sought and received information, as related to the number of workers, electrical capacity, number of stations, and the management board of the Fukushima NPPs, before the accident as they attempted to work in the FPS.

This, however, does not mean that public institutions and TEPCO did not share information with the public. All officials and TEPCO managers stated that they have disclosed information related to the Fukushima NPPs through reports, information centers, and the internet prior to the accident. Official 4 (July 27, 2017) underlines the importance of disclosure of information, in that: "*We prioritized sharing the information related to the nuclear energy and the FNA after the 2011 disaster.*" Additionally, all they stated that information related to the Fukushima NPPs and nuclear energy was provided to those who made such a request. They, however, were unable to provide any statistics relating to those who had requested information before the accident. Similarly, two out of two environmental NGOs stated that they had collected information related to nuclear waste from public institutions before the accident, while human rights NGOs had not sought nuclear energy-related information prior to the accident as, at that time, they did not have any interest in nuclear energy. Seven out of seven scholars who work on nuclear energy in Japan stated that they had collected information related to previous nuclear accidents in Japan, nuclear waste issues, and safety measures, both with and without requests, prior to the accident.

As people did not have any interest of nuclear energy, it seems unsurprising that they had not attempted to receive related information. However, risk communication is a key component of risk governance which is employed at all stages of risk governance (Renn, 2015). Risk communication is more than a simple one-way transfer of information to citizens and stakeholders. Renn (2015) argues that effective and healthy risk communication can be achieved in such a way that values, concerns, knowledge, experiences, and interpretations can be exchanged with stakeholders from disparate societal backgrounds including scientists, experts, NGOs, civil people, decision makers, and so on, which in turn supports the creation and maintenance of trust among the various actors. The fact that residents did not receive information from public institutions can be seen as poor risk communication in the instance of Fukushima.

Subtheme 2: Passive Access to Information Related to Nuclear Energy after the 2011 Accident

The passive access to the information theme refers to the idea that TEPCO or public institutions in Japan are obligated to develop national information systems and procedures that ensure the systematic and periodic disclosure of environmental information. Thirty-one out of thirty-four residents stated that they had received information regarding the Fukushima accident from reports released by TEPCO and other public institutions after the accident. Their information included radiation levels in the environment, radiation doses in food and water supplies, statistics related to health issues raised by the nuclear accident, decontamination processes in the exclusion zone, the disposition of the damaged stations, Japan's energy capacity generated from NPPs, and the number of NPPs shut down after the Fukushima accident. Residents stated that they had received this information from the following sources without prior request: (1) reports released by TEPCO, METI, the Fukushima Prefecture municipality, and the NRA; (2) brochures distributed by public institutions in supermarkets which label radiation levels in food and water supplies; (3) leaflets distributed by public institutions to residents in the public hearing and meetings which gave information about the latest news and progress with regards to radioactive contamination, nuclear energy, and the cleanup operation; (4) internet-based videos released by TEPCO and public institutions; and (5) official television broadcasts.

All TEPCO managers and central government officials stated that data related to the contamination and decontamination process is collected by experts on a regular basis. After data is analyzed, they release information as reports into the latest events, news, and developments about the FNA. It is available on their website, and people are informed about this information through the internet and face-to-face meetings. TEPCO managers and central government officials stated that they have released detailed information related to large quantities of data relating to ground conditions and earthquakes, the safety assurance system, the management of radioactive waste, exposure to radiation due to the release of radioactive materials in the course of normal operations, and accident scenarios since the 2011 disaster because "*it is the key policy to achieve transparent nuclear energy governance*" (Official 2, July 16, 2017).

Contrary to TEPCO managers and officials, all the NGOs and seven out of the seven scholars who work on environmental law stated that the government does not share clear information with the public on how energy policy is formed, who participated in the energy-making process and how the public's concern and priorities are reflected in the energy-making process. NGO 2 (June 25, 2017), for example, stated: "*we still do not know how Japan's energy policy has been formulated and who participated in the policy formulation process.*" The government's energy policy is generally known, but how it was actually formed has not previously been made clear. This means that NGOs and scholars do not find the current nuclear energy governance transparent—Scholar 3 (June 2, 2017) stated: "*Nuclear energy companies significantly influence the energy formulation process but this is not shared publicly,*" despite transparency supposedly being the key principle underlying Japan's energy policy. Article 3 of the NRA's Core Values and Principles states: "*We shall ensure transparency and appropriate information disclosure on regulations, including the decision-making process*" (NRA, 2016). There is further supporting evidence that Japan had made the effort to achieve transparency in energy governance, however, which includes the fact that all meetings between the NRA's commissioners or staff and nuclear energy companies and the outlines of the meetings' discussions are made publicly available; the NRA shares the names of the NRA attendees and the documents used in the meetings; the NRA Commissioner Meetings and other official meetings are open to the public, and the associated real-time video which is uploaded to the NRA website is avaliable in both English and Japanese; and the NRA Chairman holds a weekly press conference which is made available as a live broadcast (NRA, 2016).

All the above demonstrates that public institutions such as METI and the Ministry of the Environment share information regarding Japan's NEP, the FNA, and so on. Similarly, information about the accident, the decontamination process, risk governance, and so on, is also available on TEPCO's and the public institutions' websites. In Japan, information on radiation is published by the Japan Atomic Energy Agency, and the Japanese Ministry of the Environment publishes an associated database online. Local municipalities provide information about the radiation levels in food, water, and the environment on their websites, and which is updated on an ongoing basis. Most of this information is also available in other languages such as English or Chinese. During our fieldwork in the THUs, we also collected brochures and leaflets which provided easily understandable information about the accident, nuclear energy, radiation levels in food, water, and the environment, the decontamination process, and so on, indicating that private individuals have, or at least have potentially, received a considerable amount of information related to the accident and nuclear energy without prior request. This seems a positive step in nuclear energy governance because IEL such as the Aarhus Convention encourages countries to collect data and share the information so produced publicly and without request, as lay people may not have the time and/or money to request individual pieces of information (Anderson and Boyle, 1996). However, this does not mean that residents have enjoyed access to information. The quality of information, including its understandability, timely sharing, and so on, is as important as its disclosure, which is discussed in the following themes

Subtheme 3: Passive Access to Information Related to Nuclear Energy after the 2011 Accident

This subtheme refers to residents who have made a request to receive information related to nuclear energy in general, the FNA in particular, and environmental contamination between 2011 and 2017. Access to information implies that public institutions are obliged to provide information when a related request is made by a member of the public. Accordingly, we asked: Have you ever requested information related to nuclear energy and the nuclear accident? If so, can you please give details? Only fifteen out of thirty-four residents stated that they had previously made such a request from either the public institutions or TEPCO, and that they have managed to receive this information. Ten out of these fifteen residents were aged between fifty and ninety. They stated that they had visited information centers to receive this information because they do not have access to, or use of, the internet to receive information or they otherwise found the publicly shared information too complex to understand. Five out of fifteen residents stated that they had requested information from TEPCO by means of petition. This information included: (1) the amount of compensation that TEPCO paid for properties in the exclusion zone; (2) a schedule which shows when the cleanup process will be completed; the dispositions of their properties in the exclusion zone; TEPCO's compensation policy for those who have suffered, or are suffering, from health issues related to the Fukushima accident. It seems that TEPCO does not willingly share clear information regarding compensation because the amount of compensation is determined by individual negotiations with each family. However, we do not have any information on how much residents received or TEPCO has paid for damages to people's property and health.

TEPCO managers stated that TEPCO maintains a nuclear energy information service center, which is intended to help lay people in general and the victims of the Fukushima accident in particular to develop a better understanding of nuclear energy production, contamination, how such contamination can be cleaned up, how safety is ensured, how radiation levels are monitored and checked at the central control room, and so on. TEPCO manager 2 (August 8, 2017) defined information centers as follows: "*This centre is constructed only to provide any information related to nuclear energy to the public, and particularly concerned citizens.*" The information center employs a director, vice directors, technical section chiefs, and several public relations staff. They serve all visitors in the form of guides to help them gain a deeper understanding of the information that such individuals have requested. They also organize exhibitions and panels to inform people about aspects of nuclear energy and the contamination. The information center is open the entire year-round. All the TEPCO employees and public officials also stated that TEPCO and public institutions such as METI and the Fukushima Prefecture provide information through the internet, and via email, telephone, fax, when individuals make appropriate requests.

Scholars also raised the same point that Japan and TEPCO provide information related to the FNA when the public makes a related request because Japan has an information act that obliges the country and related institutions to collect and disclose information whether requested or otherwise. Scholars have underlined the fact that disclosure of such information became a key policy, particularly after the FNA. NGOs' main concern related to the passive access is that public institutions do not provide reliable information, particularly in relation to the risks associated with the ongoing radioactive contamination to people's lives because their intention is to gain public support for nuclear energy. However, NGOs did not state any issue in connection with the access to information when requests were made by individuals.

The literature on access to this information indicates that public institutions must share all available information when an appropriate request is made by a private individual (Anton and Shelton, 2011; Boyd, 2011). Access to this information is not purely limited to that which the public institutions disclose; certain individuals may want to obtain more information than the public institutions make generally available (Darbishire, 2010). Information is accessible subject only to a narrow system of exceptions—these restrictions can vary from country to country, but the most common exemptions are related to concerns about national security (Kravchenko, 2010). However, none of the participants stated that their information requests had been refused.

Subtheme 4: Complexity of Information

The complexity of information subtheme indicates that some residents find publicly shared information too complex to understand. Public disclosure of this information is, then, clearly not sufficient in itself to achieve access. If the information is not understandable to lay people, then access to information cannot be said to be effective (Ramkumar and Petkova, 2007; Beierle, 1998). This is particularly important in the FNA because nuclear energy is a highly technical topic that lay people may have difficulty in understanding, even, for instance, with the most basic information regarding nuclear energy (Akyuz, 2017). This was also emphasized by Resident 34 (August 29, 2017): "*Nuclear energy is a technical topic and we do not understand some information easily.*" For this reason, the literature regarding PHRs and related IEL documents carries the additional obligation that countries should provide easily understandable information to the citizens (Anderson and Boyle, 1996; Anton and Shelton, 2011).

Only the older generation raised the issue that they found publicly shared information to be overly complex. Ten out of thirty-four residents, aged between sixty and ninety, stated that they did not understand a significant proportion of the information related to nuclear energy and the accident. Their information is mostly related to the radiation and the associated health risks which include radiation levels and their impact on the environment, health, and lives; radiation doses in food and water supplies; and the decontamination of the exclusion zone. Resident 34 (August 29, 2017), who was sixty-eight years old, explained why he visits the information center: "*I do not understand the information shared with us by the public authorities. They give detailed information, but I find some of it complex. For this reason, I go to the information centre to get more detailed information or easily understandable information.*" However, twenty-four out of thirty-four residents stated that they did not have any problem understanding the shared information provided by the public institutions. Six out of seven scholars who undertook fieldwork in the THUs also underlined similar issues, namely, that the older generation, in particular, did not understand the quantitative concept of radiation; they thought that radiation is dangerous at all levels. Similarly, four out of four NGOs stated that the older generation do not understand the safety and risks concerns associated with the radiation levels for two reasons: (1) they do not typically use or rely on the internet to gain access to further information about the particular topic they have been unable to understand; and (2) they mostly live alone, which means that they do not have the opportunity to exchange information on an immediate basis (i.e., there is no one living with them with whom to share such information).

In order to mitigate this situation, TEPCO stated that they have information centers which provide easy-to-understand explanations of scientific matters to all citizens. TEPCO and the public officials stated that they do not share information alone but also provide further explanations of this information to lay people as appropriate. TEPCO 2 (June 18, 2017) explained the importance of sharing easily understandable information as follows: "*Our priority is to allow people to understand publicly shared information easily because information related to nuclear energy can be very technical.*" TEPCO employs experts in the information centers who are able to give detailed explanations of those aspects of the information that lay people generally struggle to understand. This service is provided face to face, in written form and/or orally without charge. Similarly, all the central government officials stated that disclosure of easily understandable information is a key principle of Japan's energy policy. They stated that they avoid technical and complex language when the information is released, and further that all information is explained in detail so that every member of the public can easily understand the discussion in the public meetings held in the THUs.

The fact that the older population find radiation-related information complex does not mean that complex information is necessarily shared. Even very simple and easily understandable information may be complex for any given person (Ramkumar and Petkova, 2007; Verestiuc and Tucaliuc, 2015). What is more important is that the government and TEPCO provide a service to those who find the information complex, if people do not understand it, or if they have further questions about it. Japan's NEP documents underline the importance that disclosure of easily understandable information is one of their key principles. However, this does not change the fact that the perception of the information is still complex, particularly for elderly people, which results in the fact that they have to spend time and effort to go to information centers to get easy-to-understand explanations of information that they find complex.

Subtheme 5: Reliability of Information

The reliability of information subtheme has arisen because residents stated that they do not trust some of the information regarding nuclear energy and the FNA that the public authorities share. Twenty-seven out of thirty-four residents stated that they do not find publicly shared information reliable. The information that residents do not find reliable include: (1) reports on the environmental contamination and radiation in the food and water supplies that they consume daily; (2) the possible effects on health of the accident; (3) statistics related to health incidents; and (4) safety, in terms of preventive measures for the public. Twenty out of thirty-four residents stated that the government and TEPCO are not being forthright about the risks of contamination to people's lives for the following reasons: (1) they want to increase public trust in nuclear energy; (2) they do not want to pay the appropriate compensation for the damage inflicted on residents' lives and on the environment; (3) the government does not want to lose votes of the residents due to the radiation issue; (4) TEPCO and the government want to continue the use of nuclear energy, which requires public support because local governments have the right to hold referenda on the topic of nuclear energy; and (5) the government does not want the local food market to lose value. This issue has also been raised by the environmental and human rights NGOs, namely, that the government and TEPCO does not disclose scientifically determined information about the environmental risks associated with the radiation to people's health, and the food and water supplies, in order to increase public trust in nuclear energy and to continue with the use of nuclear energy in the longer term. NGO 2 (June 25, 2017), for example, criticized the fact that "*the government and TEPCO intentionally share unreliable information to increase public support for the use of nuclear energy.*"

This, however, does not mean that the state disclosures are somehow unreliable or manipulated. All TEPCO and central government officials stated that they share only scientifically proven information with the public. TEPCO manager 1 explained that "*We only share scientifically approved information with the public, but some anti-nuclear organizations manipulate information to influence the public against nuclear energy*" (TEPCO manager 1, July 27). The government has clearly emphasized in all its NEP documents that Japan places particular importance on disclosure of scientific information to its citizens. The 2017 Basic Policy for Nuclear Energy, for example, states: "*Nuclear energy-related organizations should . . . provide accurate scientific facts and information based on objective facts*" (JAEC, 2017, p. 5). The main rationale behind this policy is to prevent, or mitigate, any fake and manipulative information related to nuclear energy, and particularly the nuclear accident, which might otherwise influence the public attitude toward nuclear energy.

The reason behind this issue is that residents receive information from different sources such as researchers, NGOs, or antinuclear organizations, each of whom have, or might have, different perceptions of the associated risks. Pronuclear bodies and opponents distribute brochures which give different interpretations of the information regarding radioactive contamination and nuclear energy. For example, the fact that radiation levels in the THUs are 4 mSv can be interpreted in many ways. While officials regard this level as safe, an antinuclear organization may see it as a serious risk to people's health because there is currently no definite scientific description of the impact of low-level exposure to radiation. How the information is interpreted and explained to the general public may affect how the receivers perceive both the information itself and the institutions from which it was derived (Ramkumar and Petkova, 2007; Verestiuc and Tucaliuc, 2015). This causes confusion among residents, where Resident 4 (June 6, 2017) stated, "*We do not know which information is reliable. While some scholars and nuclear organizations claim that radiation rate is high in terms of human safety, government officials state that there is no health risk associated with the radioactive contamination.*" However, this can be also interpreted in a positive way, in that residents receive information from both pronuclear and opposition groups, rather than just scientific information from researchers, which may improve risk communication among different groups.

Subtheme 6: Timely Sharing of Information Related to the Accident

This theme arises due to the fact that residents stated they received appropriate information at the time of the incident. According to IEL, emergency information must be provided as soon as possible after such an event (May, 2013; Boyd, 2011; Anderson and Boyle, 1996). There is, however, no universal standard as to when such information should be disclosed. Both international nuclear law and the national nuclear acts in Japan indicate a vague timeframe, with wording such as "immediately" or "as soon as possible" (Kingston, 2013a). However, the nuclear energy literature shows that exposure to even a single dose of radiation can affect the severity and type of effect on subsequent health (Christodouleas et al., 2011; Hasegawa et al., 2015; Cardis and Hatch, 2011). For this reason, it is important that disclosure of appropriate information to the public is afforded whenever the government or TEPCO obtains information related to the accident.

Twenty-five out of thirty-four residents stated that they did not receive any information about radioactive contamination over the course of the first two days after the nuclear accident in 2011. During this time, residents continued to consume local food and drink, go outside, and so on, or in other words they were allowed to continue activities that could potentially expose them to the worst of the risks from any fallout because "*the government did not inform us about the danger of radiation in food, water and air during the first few days after the accident*" (Resident 5, June 9, 2017). This was also verified by four out of the four NGOs and seven out of the seven scholars whose work involves the FNA, namely, that the government and TEPCO did not inform people who lived near the damaged stations about the potential for radioactive contamination and its risks within 24 h. As for officials, there are two different views: all the TEPCO officials stated that it was the central government's responsibility to inform local governments, who in turn should have informed the public about the accident and potential contamination. In contrast, all central government officials blamed TEPCO, in that the latter did not share information with the public institutions in a timely fashion, which is the reason why the public did not receive information on time.

There is certainly common agreement, however, that regardless of where the fault actually does lie, information related to the accident and the potential for radioactive contamination was not shared within a reasonable timeframe after the accident occurred. A nuclear accident carries a high level of risk to nearby people and the environment (Hasegawa et al., 2015; Cardis and Hatch, 2011). Certainly, as mentioned, there is no given standard as to what a reasonable timeframe might actually be, but the right to information requires that information is disclosed within a reasonable time, by the public authority (May, 2013; Boyd, 2011; Ramkumar and Petkova, 2007). It is, however, not clear what "reasonable time" refers to in any quantitative sense. From a human rights perspective, the government has a specific obligation to protect human rights such as the rights to life and health (Donnelly, 2002). The release of radioactive materials is clearly a risk to both, given that exposure to radiation for longer periods of time can increase the chances of people's health being affected (Christodouleas et al., 2011; Hasegawa et al., 2015). It seems that there are two reasons why TEPCO and the government did not share urgent information in the immediate aftermath of the accident. Firstly, there is poor communication between TEPCO, central government, and local government because they blame each other. For example, while TEPCO 2 (July 18, 2017) stated that "*the government was obliged to inform people about the accident,*" Official 3 (July 25, 2017) stated that "*TEPCO did not share any information about radioactive contamination in the first two days.*" Secondly, there is no sufficient preparedness for nuclear accident-related emergency situations, which is also underlined by central government officials confirming that they had never anticipated any accident. Official 3 (July 25, 2017) stated: "*We did not expect any accident and TEPCO was not prepared for the accident, unfortunately.*" One of the ways in which to mitigate the risk to people's lives and health would have been for TEPCO and/or the government to have informed them about the accident, and the potential for contamination, essentially as soon as any relevant information was obtained; the potential consequence here was that people continued to consume potentially contaminated food and drink, and continued to be exposed to radioactive materials. From this perspective, "reasonable time" might be considered to be the time at which the relevant institutions obtained information that suggested that people might be at (high) risk.

Theme 2: Public Participation in Decision-Making Processes

The "public participation" theme refers to the fact that residents have/have not participated in decision-making processes related to the FNA, Japan's NEP, NPPs, or the environmental contamination resulting from the nuclear accident in Fukushima, either post- or pre-2011 disaster. Thirty-four out of thirty-four residents stated that they had never attended any decision-making process regarding nuclear energy in Japan pre- or post-Fukushima. Two main reasons emerged: (1) People did not have any interest in nuclear energy and they were not invited to any public participation events. The accident, however, changed people's approach to nuclear energy in that although they might well have been more willing to attend energy-making processes, the central government had not attempted to engage the public in the decision-making process. (2) They had never been invited to participate in the decision-making process. Four subthemes emerged from the main theme, which are as follows:

Subtheme 1: A Lack of Interest in Nuclear Energy and No Invitation before the Accident

We asked whether residents had participated in any decision-making process related to nuclear energy before the 2011 disaster, and if so, how? All the residents questioned stated that they had not attended any such decision-making process prior to the accident, for which they gave two main reasons. Firstly, they all stated that they did not previously have any interest in nuclear energy-related activities, decisions, policy, and so on, prior to the accident; secondly, they noted that they were never at any point invited to participate in any decision-making process related to nuclear energy by TEPCO, or the central or local governments. The latter was confirmed by TEPCO officials, specifically that TEPCO did not invite local people to participate in the decision-making process or risks assessment because the company did not have any legal obligation to do so when the Fukushima NPPs were constructed in 1967. Four out of the four central government officials noted that Japan had engaged the public in the decision-making process before the 2011 disaster, but also that the participants were selected randomly. They did not have any data about whether residents were invited to public participation in the decision-making process before the Fukushima event.

NGOs and scholars verified that Japan attempted to engage the public in the decision-making process before the FNA through round table, surveys, and questionnaires all over the country after, in particular, the Tokaimura nuclear accident in 1999. However, they stated that the public did not play any active role in policy formulation process because the pronuclear government and nuclear industry did not want to share such power with the public. In other words, public participation was sought for purely tokenistic reasons so as to increase public support for nuclear energy. Central government officials also noted that Japan had engaged the public in the decision-making process before the 2011 disaster, but also that the participants were selected randomly. They did not have any data about whether residents were invited to participate in the public side of the decision-making process before the Fukushima event. However, the difference between officials and NGOs is that while officials claimed that people's concerns and priorities were reflected in the energy-making process, NGOs stated that public did not make any meaningful contribution to the formulation of the nuclear energy process.

Normally, after permission is obtained from local governments, nuclear energy companies begin the process of risks assessment under the Environmental Impact Assessment Act (Elliott, 2013). During the formal environmental assessment stage, public comments are sought. The nuclear energy company must then hold a public hearing (NRA, 2015). All public comments are submitted to METI to gain approval for their projects (Kingston, 2012). However, the Environmental Impact Assessment Act was only enacted in 1993, and the Fukushima NPPs' construction began in 1967; there was no law obliging TEPCO to communicate with local people about the risk assessment or the construction of the Fukushima NPPs at that time (Elliott, 2013).

Subtheme 2: People's Interest in Public Participation in Decision-Making Process Related to the Nuclear Energy after the Fukushima Accident

This research found that, unsurprisingly, the Fukushima NPPs accident altered people's interest in nuclear energy that they have been considerably more willing to participate in any decision-making processes since the 2011 disaster. Twenty-five out of thirty-four residents stated that they were interested in attending nuclear energy-related discussions in order to influence Japan's energy policy. The main reason behind this change was clear from the fact that all residents stated that nuclear energy could affect the quality of their lives, as evidenced by the Fukushima accident, and that the government had failed to govern energy sector, again as evidenced by the accident. Resident 19 (July 18, 2017) stated: "*I was not interested in nuclear energy before the accident, but the accident changed my approach to nuclear energy. I have really wanted to be a part of the decision-making process since the accident happened.*" Residents, however, raised a number of issues and obstacles which affected people's interest in public participation via decision-making processes.

Firstly, fourteen out of seventeen female participants were housewives, who stated that it is not appropriate in Japanese culture for women to participate in any decision-making process because men, who are considered the heads of the household in the Japanese family system, represent the entire family. They stated that this is not a women's responsibility, and rather it is the oldest men in the family who are responsible for talking about and making these decisions. Japan is indeed a very male-dominated society. Men head households and manage their families' estates; this stereotype is particularly prevalent in rural Japan and it should be noted that many residents used to live in the villages or countryside before the accident (Gerteis, 2009). Resident 24 (July 20, 2017) underlines the same issue that "*We are equal in Japan, but our husbands or fathers rule the family. For this reason, it is more appropriate that men attend meetings with the TEPCO and the officials.*" This shows that local values appear to be an obstacle to women participating in the nuclear debate. There is, however, no legal restriction in this regard, as all rights are the entitlement of all people, regardless of their gender. Notably, no male participant stated that women were prohibited from participating in decision-making processes.

Secondly, sixteen out of thirty-four residents who were employed stated that they were unable to regularly participate in any decision-making process because they work hard (another aspect of Japanese culture) and consequently did not have time for public participation despite any interest they might have in the nuclear energy matter. Resident 8 (June 8, 2017) explained the importance of the culture of overwork in Japan, in that: "*We prefer to spend time on our careers to time on any social or political events in Japan.*" However, eight out of thirty-four residents were retired, and stated that they had sufficient time, and were indeed willing to participate in activities regarding the FNA and nuclear energy accordingly. None of the housewives raised overwork as an obstacle to their public participation; this was limited to those who were employed. As implied above, overwork is a cultural value in Japan. People work until late and will continue to work over the weekend. Every year, hundreds of people in Japan die or suffer from serious illnesses due to overwork, and which can be translated literally as Karōshi (過労死) in Japanese (Ishiyama and Kitayama, 1994).

Thirdly, residents believed that the pronuclear party and the nuclear industry are the only stakeholders determining the future of nuclear energy in Japan. The "nuclear village" is a solid triangle of bureaucracy, industry, and academia (Cotton, 2015). Twenty-three out of thirty-four residents stated that the coalition between the pronuclear LPD party and the nuclear industry represents a barrier to concerned people being able to influence Japan's NEP despite public opposition to nuclear energy being high, as this could result in the closure of all NPPs. This issue was mostly raised by younger participants, typically those aged between eighteen and forty years. Resident 9 (June 8, 2017) explained why he is pessimistic about public participation: "*I would like to be a part of public participation, but nuclear energy companies do not let us influence nuclear energy policy in Japan.*" This issue was also emphasized by all the NGOs that nuclear energy-related decisions are made by the cooperation between the pronuclear government LDP and the nuclear industry, and which does not want to share power with the public.

The nuclear village could well be an obstacle to public participation, but there is no tangible evidence to actually prove this. It is a fact that both the LPD and the nuclear industry share a common commitment to nuclear energy. Secret meetings between the government regulators and bureaucrats, and the representatives of nuclear energy companies between 2012 and 2015, have recently been disclosed by the media (Kingston, 2012); there is, however, no information about what was discussed during these meetings, or how they might have shaped Japan's energy policy. The NRA shares the names of the NRA attendees and the documents used in these meetings, and indeed the NRA Commissioner Meetings and other official meetings are open to public participation (NRA, 2015).

Subtheme 3: Public Participation in Decision-Making Process—Tokenism

This theme considers the fact that people have participated in the decision-making process, but have been unable to make any meaningful contribution to the associated decisions. According to Arnstein (1969), informing, consultation, and placation occupy the middle rungs of the participation ladder and are thus considered "tokenism," which refers to the idea that while some roles are given to the people, but they do not have any real influence on any decisions taken. This research found evidence that informing is the only public participation type that individuals have experienced in the FNA. TEPCO and officials organize public meeting to inform people about the nuclear accident, decontamination process, projects regarding the damaged stations, the situation of THUs, measures for dealing with radioactive contamination, compensation regarding the impacts of the accident on people's property, and so on. Thirty-four out of thirty-four residents stated that they have participated in public meetings organized by METI. According to the residents, the concept of these public meetings is as follows: (1) TEPCO managers and central government officials announce a meeting one month in advance; (2) all residents aged over eighteen are invited to these meetings; (3) TEPCO managers and central government officials explain the latest news, plans, and so on; (4)

residents explain their thoughts and share their concerns with the officials; and (5) residents ask questions about nuclear energy and the contamination.

All NGOs agreed that these meetings are not of any particular influence on the decision-making process. They noted that public meetings do not play any essential role in Japan's energy policy formulation process because "*the nuclear village does not want share power with the public with regards to nuclear energy in Japan*" (NGO3, July 26, 2017). Residents only express their fears and concerns about the risks and listen to TEPCO's and the central government's information. Environmental NGOs stated that public meetings aim to alter people's opinions by manipulating public opinion itself in the attempt to bring it into line with the government's goals. For example, NGOs stated that officials and TEPCO managers inform the public that there are no risks associated with the contamination on people's lives. However, there is approximately 1.09 million tons of highly radioactive water currently stored in 900 tanks in Fukushima seeping into the ocean, which is contaminating the water table for the whole region. NGOs stated that TEPCO and other officials never seem to mention the risks associated with the contamination to the environment and people's lives. In contrast, all central government officials stated that they have organized public meetings with residents to hear their comments on relevant issues, and to identify people's concerns about nuclear energy and the accident. They note all comments and submit them to the NRA, which is responsible for Japan's energy policy. When decisions are made, residents' comments, concerns, and priorities are taken into consideration.

Officials imply that residents influence nuclear energy-related decisions, but that public meetings are related to risk communication between the public, the energy company, and the central government with regards to the Fukushima accident rather than public participation in the decision-making process about Japan's ongoing energy policy. Public meetings are limited to discussion into the radioactive contamination, the decontamination process and environmental safety in the region, rather than Japan's energy policy. This was emphasized by Official 2 (July 16, 2017), in that: "*the public does not have any right to get involved in decisions related to nuclear energy in Japan, but public concerns are reflected in energy policy.*" This shows that officials and residents have different understandings of the role of the public in the decision-making process because while officials state that public concerns and priorities are reflected in decisions related to nuclear energy, residents believe that they do not have any influence on nuclear energy-related decisions.

In other words, there is a two-way flow of information. Residents do not have any power of negotiation during the informing process because there is no channel for feedback. However, residents have the opportunity to raise their voices about any concerns related to nuclear energy in these meetings. Additionally, meetings are generally about aspects of governance rather than policy. Public meetings are limited to discussion about the radioactive contamination, the decontamination process and environmental safety in the region, rather than Japan's energy policy. This means that the residents have no influence at the policy level. The main purpose of these meeting is only to inform people about the above topics rather than incorporating residents' values, risk perception, or concerns into decision and policy; for this reason, this participation remains purely tokenistic; it does not allow for any true or meaningful participation.

Subtheme 4: Referendum Demand

The referendum subtheme refers to residents' demand for a referendum on NEP from the government. In order to better understand residents' thoughts about public participation in the energy formulation process, we asked: "*What do you think about nuclear energy? Are you against or for nuclear energy?*" Thirty-four out of thirty-four residents raised objections to nuclear energy and twenty-six out of thirty-four residents stated whether they wanted a referendum over the usage of nuclear energy in order to vote for or against having NPPs in their neighborhoods. However, this does not mean that people are against NPPs or nuclear energy in Japan; the country has fifty-four NPPs, but only twenty-six out of thirty-four residents raised objections to NPPs in their neighborhoods, namely, Fukushima, while eight out of thirty-four residents demanded a referendum about the existence of NPPs in the country. This shows that there is no consensus among residents over whether Japan should continue with nuclear energy at the national level. As stated by Resident 9 (June 8, 2017): "*I do not want NPPs in Fukushima. However, I do not mind whether Japan has NPPs in other prefectures.*"

We asked TEPCO managers and central government officials, NGOs, and scholars a direct question about what they thought about the possibility of a referendum over the use of nuclear energy in the country. While all TEPCO managers stated that they did not have any idea about such a referendum because this would be a political process between the government and the public, central government officials stated that a referendum is not on the current government's agenda. Four out of four officials stated that listening to people's concerns and priorities can be far more effective than a yes/no referendum in designing Japan's energy policy. Official 2 (July 16, 2017), for example, stated: "*People have concerns about nuclear energy because they are angry about the nuclear accident, but we will not let them vote for or against NPPs purely in anger.*" All the NGOs and scholars who work on environmental law also stated that voting in a referendum on the use of nuclear energy has been a common demand among citizens in Fukushima since the 2011 disaster, but is not on the government agenda because public opposition to nuclear energy is high due to the 2011 disaster and its likely consequences.

The referendum does not seem likely in the short term because all the officials stated that Japan needs to continue using nuclear energy in the short term to meet its energy demands, even though this is not what the public apparently wants. Public opposition is high due to the nuclear disaster and ongoing radioactive contamination in the region, but Japan needs nuclear energy to achieve its 3Es plus S policy, and ultimately this appears to be the reason why the government does not currently want to let people become directly involved in the energy policy-making process.

Theme 3: Access to the Court

The "Access to the Court" theme refers to whether people have access to the courts for any issues in relation to nuclear energy, the FNA, radioactive contamination and its impact on their lives, and NEP in Japan. No resident has applied to the court for human rights issues in relation to the environment but twenty out of thirty-four residents stated that they have applied to the court for the nuclear accident and environmental contamination-related issues. A number of subthemes were generated, which are as follows:

Subtheme 1: Judicial Review with Respect to Environmental Decision Making

This subtheme refers to whether residents, NGOs, or other stakeholders have applied to administrative review or another independent and impartial body to review any decision in relation to environmental and nuclear energy-related issues, and other issues in connection with access to information between 2011 and 2017 before the court. According to the EHRs theory, states are required to ensure access to a review procedure in connection with decisions and information made by the public authorities before the court because a court review can be lengthy and expensive, and by contrast environmental information and decisions are generally needed quickly (Anderson and Boyle, 1996; Anton and Shelton, 2011). This right ensures that people enjoy the right to public participation in decision-making processes and have the right to access appropriate information (Boyd, 2011).

Thirty-four out of thirty-four residents stated that they have never applied to any independent or impartial body to review environmental and/or nuclear energy-related decisions, and issues in connection with access to information pre- and post-Fukushima because their main concerns are related to environmental contamination, the decontamination process in the exclusion zone, compensation for damage to their properties, and radiation-related risks rather than nuclear energy-related decisions. Resident 7 (June 7, 2017) explained the reason that "*our priority is to get compensation for damages to our property and lives. We do not have time to review nuclear energy-related decisions.*" Additionally, none of the residents' information requests have been refused and residents have never participated in any decision-making process. For these reasons, they have never applied to any independent body to review any nuclear energy-related decisions or for refusal to provide the information they have requested.

No issue in connection with judicial review with respect to environmental decision making has been raised by any participants because, as officials and scholars stated, Japan has independent courts which review the lawfulness of the decision-making process when people or civil society organizations apply to them.

Subtheme 2: Judicial Redress with Respect to the Impact of Environmental Decisions

This subtheme refers to whether residents have applied to either courts or tribunals to seek remedy for the impact of the FNA on their lives, health, property, environment, living conditions, and standards between 2011 and 2017 in Fukushima. Environmental harm and benefit have not been shared equally among the members of society, which is referred to as distributive environmental injustice (Holifield, 2001; Cotton, 2017). Accordingly, we asked residents whether they had ever applied to any court to seek remedy for the impact of the FNA on their lives, health, property, or the environment.

Twenty-four out of thirty-four residents stated that they have applied to the Fukushima District Court to receive compensation for damages to their properties in the exclusion zone and for the poor living conditions in the THUs. The exclusion zone consists of three parts: (1) the "Difficult to return zone," which has more than 50 mSv annual integrated doses; (2) the "Restricted residence zone," which has 20–50 mSv annual integrated doses; and (3) the "Evacuation order cancellation preparation zone," which has 20 mSv annual integrated doses (Figueroa, 2013; Kingston, 2012; Hultquist and Cervone; 2018). All these residents have their proper houses in the restricted residence zone, and to which they do not have access to due to the radioactive contamination. Typical among these people is that they have children. Residents state that the THUs are generally too small for a family with children, are of poor quality, and isolated from the city centers. They stated that they found the levels of compensation to be too low and that the THUs are uncomfortable to live in. They want to buy houses in different cities with compensation for their properties in the exclusion zone. Similarly, all human rights and environmental NGOs, and six out of fifteen scholars (the other nine scholars stated that they had no information about compensation as they did not have any direct academic interest in this issue) stated that the compensation that TEPCO has paid out is low for accommodation and residents suffer from poor living conditions in the THUs.

In contrast, all the TEPCO and government officials stated that they had paid sufficient compensation to all residents who had been prevented from having continued access to their properties and were providing free housing for victims of the nuclear accident. They stated that typical Japanese houses are in any case small and made of wood, as are the THUs. Indeed, during our fieldwork, we witnessed the fact that Japanese houses are indeed tiny, typically being built on plots of land that are less than 50 m^2, even in Tokyo, Kyoto, and Osaka. They did not indicate the amount of the compensation that they paid, and TEPCO recommends that compensation be negotiated on an individual basis, which means each person or family may receive a unique amount of compensation because payments vary depending on the amount of radiation recorded in a particular area. As residents do not pay any rent for the THUs, TEPCO and the government officials stated that the compensation currently offered for damaged properties is reasonable. The lack of known standards regarding the amount of compensation suggests that there is a lack of transparency regarding compensation in risk governance.

While all residents had their own homes before the accident, they now live in small and isolated public houses due to the contamination. We visited more than ten THUs; they do not have social areas for children and families, are quite small for a family, and do not have public transport links nearby. More importantly, they are close to the exclusion zone, which makes people worried about being exposed to radioactive materials in the area. It seems that they have sufficient reason not to be happy with their current situation. As for compensation, TEPCO is paying 8,000 yen (£54) compensation per night until victims are able to restart their lives in their former hometowns (TEPCO, 2018). The amount is, however, not low because the average price for a hotel room for a single family varied between £40 and £100 in Fukushima when the researcher was in the region in 2017. People may find this compensation low as a hotel room or a THU is not in any way similar to their previous houses in the exclusion zone. All this discussion indicates that residents have access to the courts with regards to the environmental impact of the FNA on their lives. It also shows that people have been exercising their right to redress, as guaranteed by Constitution of Japan. However, this does not mean that people have been given sufficient compensation for the environmental damages to their lives, health, and property, as they consider it to be too low.

Subtheme 3: Barriers to Access to the Court and Socially Fair Results

The "Barriers to Access to the Court and Social Fair Results" subtheme refers to the fact that all members of the society must have equal accessibility to the courts and fair results in relation to nuclear energy issues, the FNA, radioactive contamination and its impacts on people's lives. According to the EHRs, there are a number of obstacles to the enjoyment of the right to justice, including geographical barriers, the high expense of a lawsuit, lack of education or knowledge about the legal system, lack of access to legal aid, distance, and so on (Armeni, 2016; Oliver, 2013; Anderson and Boyle, 1996). Accordingly, we asked: Do they have any social, economic, or political barriers to access to justice for the FNA? If so, what are these barriers?

None of the residents has raised any issues in connection with access to the court in the matter of the FNA. They also stated that volunteer lawyers and some local NGOs informed them about legal aid and have encouraged them to apply to the courts for the environmental damages to their lives and properties. All the residents stated that they have not sought or received any assistance from TEPCO or government officials in terms of access to the courts because "*none of the residents trust TEPCO or government officials about the FNA*" (Resident 15, July 5, 2017). All NGOs stated that the victims of the accident have been given sufficient opportunity to apply to the courts for the environmental damage to their lives, health, properties, and so on. Similarly, seven out of the seven scholars working on environmental and human rights law in Japan stated that Japan provides legal aid to all the victims to ensure that they have access to the courts and that there are no legal barriers to access to the courts. All government officials and TEPCO managers stated that they have paid all the compensation ordered by the courts due to loss of property and poor living conditions (in the THUs) in relation to the accident.

As for fair results, all plaintiff residents, scholars, and NGOs agreed that the Japanese courts are independent of both government institutions and TEPCO and, further, they provide fair results in connection with the impact of the FNA on people's lives, health, and environment. Twenty out of twenty-five residents stated that they had won lawsuits against TEPCO and the government, while four residents are still awaiting court decisions. None of the residents wanted to state the amount of compensation that they had been awarded. This can be interpreted in such a way that the amount of compensation that the government and TEPCO have offered is insufficient to compensate for the impacts of the accident on people's lives and property, as the courts ruled TEPCO to pay extra compensation. This all shows that the courts are equally accessible to all individuals and that it produces individual and socially fair results, which is a key aspect of access to justice (Armeni, 2016; Oliver, 2013).

DISCUSSION

Only a very limited amount of research has considered PRs in the nuclear energy-making process and the Fukushima accident in Japan (Aldrich, 2013; Vivoda, 2012). Such research has, for the most part, focused only on one aspect of PHRs such as the right to public participation in the decision-making process, or access to information (White, 2015). There has been no comprehensive research which has simultaneously investigated the three pillars of PHRs including the right to public participation in decision-making processes, the right to information, and the right to justice in the FNA. My research has investigated Fukushima residents' experiences of PHRs in the Fukushima matter between 2011 and 2017, and has found key issues and needs that appear to be obstacles to the enjoyment of PHRs by individuals. The discussion is formed around three subtitles, including (1) public participation in the decision-making process, (2) access to information, and (3) access to justice by critically comparing my findings with previous research into the topic.

Access to Information

This research has found that residents have both active and passive access to information regarding the FNA and nuclear energy in Japan. Public authorities and TEPCO share information related to nuclear energy. However, this does not imply that residents have enjoyed effective access to this information. This research has identified some key issues in relation to access to the information. Firstly, my data shows that residents do not find publicly shared information reliable. Literally speaking, reliability is the extent to which a person can trust the data itself and its source (Cramer, 2009; Darbishire, 2010). Reliable information refers to being dependable, trustworthy, unfailing (Pierce, 2008). However, my findings show that residents do not trust information related to, particularly, radioactive contamination, safety measures, and risks associated with the nuclear accident to their lives. There has been some research with similar results (Thomson et al., 2012; Murakami et al., 2016). According to comprehensive empirical research conducted by Tateno and Yokoyama (2013), only 111 out of 1,793 residents and 46 out of 1,793 residents in Fukushima trust information by the government and nuclear energy companies, respectively. Similarly, Murakami et al. (2016) found that many people still refuse to eat local food, even though the government shares information with the public that the food supply is safe. Accordingly, the public tends to generate and analyze their own data due to this loss of belief in the reliability of publicly shared information (Thomson et al., 2012; Murakami et al., 2016). They have the citizen-volunteered radiation data which measures collection and publication of data on radiation exposure, which is independent of official statements, free of comments, and thus politically neutral (Aldrich, 2013).

The fact that residents do not trust information does not indicate that public institutions share unreliable information. Contrary to my findings, Japan's energy policy documents indicate that disclosure of reliable and scientifically proven information related to nuclear energy and the radioactive contamination is a key principle in nuclear energy governance as explained in chapter 3. For example, the government adopted the 2017 Basic Policy for Nuclear Energy which develops a more systematic approach to access to information and encourages state institutions to set up information systems based on scientifically accurate information and objective facts. The policy document states: "*Nuclear energy-related organizations should . . . provide accurate scientific facts and information based on objective facts*" (JAEC, 2017, p. 5). However, disclosure of scientifically approved information is not sufficient to gain public trust in information because the literature on the EHRs theory shows that reliability of information is dependent on the reliability of the associated information sources (Anderson and Boyle, 1996; Pierce, 2008; Hajjar Leib, 2017). This means that if people do not find the information sources reliable, they tend to distrust the information (Darbishire, 2010). Firstly, my research shows that residents do not trust the information's source because they believe that TEPCO and the government have failed to control radioactive contamination and to protect the citizens from radiation, which has also been underlined by previous research that public trust and confidence in such decisions were easily shaken by poor risk governance during and even after the nuclear accident (Kingston, 2013b; Figueroa, 2013). Such examples include: (1) when the accident happened the government announced that "*The radioactive fallout does not have any immediate health effects*" making the public doubt the government's reassurances (Tateno and Yokoyama, 2013, p. 2); (2) the Japanese government did not publish any information on food safety until two weeks after the disaster (Reiher, 2017); (3) the Nuclear Regulatory Agency announced in June 2016 that there was some radioactive contamination from the damaged stations in the tap water of several prefectures, although the water quality has been declared perfectly adequate, is monitored by the public authorities, and appropriate information has been shared with the public since 2011 (Yontrarak, 2018). These and similar events have eroded public trust in the government, which in turn triggers distrust in official information.

By contrast, Tateno and Yokoyama (2013) and Murakami et al. (2016) note that people do not find information reliable due to scientific uncertainty about radiation and different risk perception. While public institutions take measures in accordance with the national standards, people perceive the measures and associated risks to be unsafe. The lack of acknowledgment of socially acceptable risk or ignoring subjective risk on the part of public institutions triggers this issue. Indeed, the health effects of low doses of radiation are somewhat uncertain (Takahashi, 2014). While the International Commission on Radiological Protection published a report recommending that humans control annual individual exposure to radiation to 1 mSv/yr, the Japanese government declared that a radiation dose between 1 and 20 mSv is safe, which is an objective risk to people's lives and health (Sugimoto et al., 2014; Monzen et al., 2011; Eisler, 2013). This risk is acceptable for public institutions; however, the risks associated with the radiation levels are perceived in a subjective manner by the public because the risk perception literature indicates that risk perception is relative, as one's particular perspective of a hazard will depend on personal values, beliefs, and the worldviews of members of the public (Sjöberg et al., 2004; Slovic, 2000; Mah et al., 2014). For example, if people smoke, have an unhealthy lifestyle, and do not exercise, low-level radiation doses may have more impact on such an individual's health than those who have a healthy lifestyle, or for instance if the person has a family member suffering from thyroid cancer, they can perceive a radiation dose of between 1 and 20 mSv to be dangerous even though the government has declared it safe (Jargin, 2011). Therefore, the uncertainty and different risk perception is another reason of why people do not find publicly shared information reliable.

The second main finding relates to the timely sharing of information. The literature on EHRs theory shows that access to information can be effective when this information, about activities that affect the environment, is delivered in a timely fashion (May, 2013; Xiao et al., 2017). When the information is significant, particularly in emergency situations, it is called emergency information by the Aarhus Convention which states that: "*public authorities to disseminate immediately all information. . .*" (EU, 2001). This study has found that the government and TEPCO did not share emergency information related to the radiation with the public in the first (critical) few days after the accident. This finding is consistent with previous research which indicates that residents did not receive information regarding contamination to the environment, food, and water resources in good time after the accident happened, which resulted in the issue of victims potentially being exposed to radiation for longer than necessary (Reiher, 2017; Hasegawa et al., 2015). The uncertainty derives from the fact that neither nuclear energy nor environmental law obliges the government to a specific time in which pertinent information must be shared with the public. Hallo (1996) argued that public institutions must share emergency information whenever they obtain appropriate and relevant data. However, raw data can be too technical for lay people to understand (Cramer, 2009; Darbishire, 2010). For this reason, institutions may need more time to analyze data and to share it in an appropriate manner with the public. Information can be shared whenever it is obtained as the government has an obligation to protect people's rights to life and health because the empirical research on radioactive contamination indicates that being exposed to the radiation for even a short time can have serious health impacts (Christodouleas et al., 2011; Hasegawa et al., 2015; Cardis and Hatch, 2011).

Public Participation in the Decision-Making Process

My research has identified a number of key issues in relation to public participation in the decision-making process in Japan. Firstly, my research has found evidence that residents did not have the opportunity, or indeed the invitation, to participate in the decision-making process related to the Fukushima NPPs or NEP in Japan in general before the 2011 accident. However, this does not mean that there has been no public participation prior to the disaster. Contrary to my findings, other empirical research into public participation in nuclear energy-related decisions in Japan indicates that Japan had indeed engaged the public in the energy formulation process at the national and local levels before the Fukushima accident (White, 2015). Such examples from the available research include round-table conferences on NEP between 1996 and 1999, public comments and public hearings on the draft long-term plan for nuclear energy in 2000 and 2004, referenda in 1996 and 2001, and a deliberative poll in 2012 (White, 2015; Hayao, 2014). There are two reasons behind the difference between my findings and the related literature. Firstly, Fukushima is an older NPP that was constructed in 1967 (Vivoda, 2012). Many of the residents we interviewed moved to Fukushima after the 1960s. Secondly, people were randomly selected for the public participation held by the government before the 2011 accident at the national level, and a limited number of participants attended these participations (White, 2015). PHRs literature suggests that those who are affected, or will likely be affected, by nuclear energy should participate in any decision-making process in the attempt to ensure effective governance (Anderson and Boyle, 1996; Anton and Shelton, 2011; Boyd, 2011). However, although Fukushima is one of the largest nuclear plants in Japan, none of the residents have been invited to any of the public participation held before the 2011 disaster. This shows that public participation was not common or indeed extensive at the national level before the FNA.

This research found that the accident has changed the way that people consider nuclear power; understandably, they are much more willing to interact with the government with regards to the energy formulation process to a much greater extent. Most residents stated that they would like to contribute to the decision-making process related to nuclear energy. There are a few examples of empirical research reporting similar results (Kingston, 2013a). Aldrich (2013), for example, notes that many of Japan's citizens now wish to participate in the decision-making process related to nuclear energy since the FNA. Similarly, Nakamura (2017) conducted empirical research which showed that the accident has changed people's views, and that they are now more willing to be a part of the energy formulation process in Japan. However, Nakamura (2017) noted that people's willingness to participate in the decision-making process has decreased to 40% since the 2011 disaster due to a lack of political will by the government and social hesitation. This is indeed consistent with my findings that residents believe that the "nuclear village" does not want to share power with the public with regards to nuclear energy, mitigating people's willingness to attend the nuclear energy decision-making process.

There are a large number of studies that claim the so-called nuclear village and its institutionally embedded power is attempting to determine the path of Japan's energy policy as discussed above (Feldhoff, 2018; Hasegawa, 2014; DeWit, 2012; Cotton, 2015). Shadrina (2012, p. 76) emphasizes the same issue in stating that "old habits die hard," meaning that the Fukushima accident did not change the issue, namely, that a solid triangle of bureaucracy, industry, and academia, the "nuclear village," has kept control over NEP. The role of the nuclear village on Japan's energy policy has also been emphasized by Prime Minister Kan. He publicly announced in 2011 that he intended to phase out nuclear power completely. Since leaving office, Kan has criticized the "nuclear village" responsible for Japan's energy policy (Otsuki, 2016). This, however, does not go beyond claims. Japan's energy policy documents clearly indicate that all decisions are taken by the public authorities established by the law in a transparent way, and where transparency is one of the key principles in Japan's energy policy (NRA, 2016), as indicated in chapter 3. However, Kingston (2013b) takes this argument further in claiming that the NRA is headed by prominent pronuclear experts. This still does not prove that the nuclear village controls the decision-making process because being pronuclear does not necessarily mean that decisions are taken with the cooperation of the nuclear industry. Additionally, the nuclear village claims are not consistent with the NRA's policy in practice, because they state that Japan's energy policy is targeting a decreased dependency on nuclear energy while increasing renewable energy (NRA, 2016), as discussed in chapter 3. The Japanese government forecast a significant increase in the share of NPPs, up to 50% of the electricity supply in the 2010 Basic Energy Plan through the creation of fourteen new nuclear reactors by 2030, but the FNA changed its longer-term energy strategy in a way that NPPs will be reduced from 30% to 22% of the country's total electricity production by the same year (IAE, 2016). This shows that the nuclear energy-related decision did not favor the nuclear companies, meaning that the nuclear village may not be the main reason why the public does not participate in the decision-making process.

This research has found evidence that residents participate in the decision-making process referred to as informing by the typology defined by Arnstein (1969). My TA indicates that there is a one-way flow of information from powerholders to residents in Fukushima. The powerholders engage residents in the decision-making processes related to nuclear energy for tokenistic reasons, meaning that people do not have any real influence on energy policy. Powerholders aim to only inform people about topics regarding aspects of governance such as decontamination and radioactive contamination. This is also in line with previous research into public participation in decision-making processes about nuclear energy in Japan—there has been very limited research into public participation in Fukushima. For example, White (2015) provides empirical data in his doctoral dissertation regarding this topic, in that residents do not have any meaningful participation in the decision-making process, even though meetings are frequently held between citizens and powerholders. According to this research, a lack of political will on the part of the government is the main reason why people are not becoming involved in the decision-making process. Ultimately, then, people do not make any meaningful contribution to public participation in the decision-making process in Fukushima.

Lastly, my research has found that residents support a referendum which has not been discussed by previous research into the subject. The main reason is that public opposition to nuclear energy has been high since the 2011 disaster, which is in line with the literature (Joskow and Parsons, 2012; Fam et al., 2014). The central government has conducted some surveys, which demonstrate that public opposition to nuclear energy is running at around 80% (Kingston, 2012). The Prime Minister's Cabinet Office conducted surveys in 2009 and 2011, which showed a decrease from 60% to 2% of participants supporting the increased use of nuclear energy between these two years (Aldrich, 2013). This shows that a referendum over the usage of nuclear energy was almost certainly in favor of their discontinued use. However, the policy analysis in chapter 3 indicates that Japan's current energy policy is based on the 3Es plus S policy in the short and medium term, which was intended to be achieved through nuclear energy and renewable energy. For this reason, a referendum would threaten Japan's energy policy because the current government plans to continue to use nuclear energy until the 2030s (Vivoda, 2012).

Access to Justice

This study found that residents have applied to local courts for compensation for damages to their properties and poor living conditions in the THUs. Having access to the court, however, does not mean that environmental justice is achieved. Processes in the court's jurisprudence should be designed in such a way as to lead to a fair outcome (Anton and Shelton, 2011; Hajjar Leib, 2017). How courts take decisions independently and fairly is also important (Anderson and Boyle, 1996) because the government, or TEPCO, may influence the courts' decisions. No issue in relation to this is raised by either residents or other stakeholders such as NGOs and scholars. Additionally, the decision of the courts, which order TEPCO and the government, indicated that the courts are independent in the Fukushima matter, which enables victims of the accident to achieve environmental justice. TEPCO and the public institutions respect the courts' decisions and they paid all compensation that the courts ordered between 2011 and 2017. All these results are consistent with extended literature which shows that residents seek remedy for environmental damages to their property (Feldman, 2012; Shrader-Frechette, 2012; Osaka, 2012; McNeill, 2012).

Chapter 5

Discussion

INTRODUCTION

The main aim of this section is to interpret and discuss the importance of the major findings of this book in the light of the previous research regarding the FNA and EHRs. It discusses how the new insights into the FNA event and EHRs theory have emerged as a result of this research by linking my evidence to the previous research. It also discusses how the research questions have been answered and the research objectives have been achieved. Lastly, it acknowledges a number of limitations to the research.

This chapter focuses on two main areas. It firstly argues the extent to which the FNA is a human rights issue as per its effect on the environment, which is the key research question addressed in the empirical chapters. My empirical research shows that the rights to life, health, and property have been affected by the ongoing contamination in Fukushima, but there has been no detailed research investigating human rights issues in connection with the environment in Fukushima. Accordingly, this section explores how the FNA has caused human rights issues in connection with the environment. It then investigates the cross-cutting issues between Japan's human rights and energy policy, as nuclear energy has affected the human rights of residents. There have been a number of research efforts that have considered the integration of human rights and climate change policy worldwide, as the environmental impacts of climate change on people can be considered a human rights violation, but there has been no research that explores the linkage between human rights and energy policy in Japan. Accordingly, we explore how Japan's human rights and energy policy are related to each other and what cross-cutting issues Japan's human rights and energy policy share. Lastly, this study brings this empirical evidence into a theoretical discussion into whether there is a need for a distinct right to the environment. This helps us achieve the first research objective, namely, that of exploring the relationship between the enjoyment of human rights and the contaminated environment in Fukushima. The second main finding is that the government has violated PRs with regard to the FNA, though this issue has previously been explored by various research efforts. What is new, however, is that we found that this violation of PRs triggered an associated violation of SHRs. This section investigates the relationship between substantive rights and PRs in Fukushima by addressing the second main research question: "What are the advantages and disadvantages of framing nuclear energy policy and its possible impacts, in the context of environmental human rights?" In doing so, we aim to achieve the second research objective, which attempts to investigate the advantages and disadvantages of the recognition of EHRs, particularly PRs, in relation to nuclear energy matters in Fukushima. By doing so, the main contributions of this book to the literature and why they are important will be explored.

MAJOR FINDINGS AND THEIR CONTRIBUTIONS TO THE EXISTING LITERATURE

This research achieved two major findings which answer two of the overarching research questions of this book. The first of these is: "How does nuclear energy relate to environmental human rights in Japan?" This helps us to better understand the relationship between the quality of the environment and the enjoyment of human rights. The second main research question is: "What are the advantages and disadvantages of developing a nuclear energy policy that enables the citizen to enjoy their environmental human rights?" The main research objective here is to explore the connection between PRs and substantive rights. The following section argues the importance of these findings and their contributions to the literature, respectively, in the light of the previous research regarding the FNA and EHRs.

Main Finding 1: The Relationship between Human Rights, the Environment, and Nuclear Energy in Japan

The following sections consider the first research question through three steps, and accordingly, this section consists of three parts. It first discusses the cross-cutting issues between Japan's energy and HRP. It then focuses that the environmental impact of the FNA is in violation of the rights to life, health, and property in Fukushima. Lastly, it critically interprets what this finding means in terms of EHRs theory, which argues as to whether there is a need for a new right to the environment or whether existing human rights already require a safe environment as a prerequisite.

Cross-Cutting Issues between Japan's Human Rights and NEP

The policy analysis in chapter 3 has for the first time identified that environmental safety is a cross-cutting issue between Japan's HRP and NEP. However, neither Japan's HRP nor Japan's NEP references each other in any way. This is not a situation that is unique to Japan because the relationship between nuclear energy and human rights has not as yet been documented at any intergovernmental and governmental level. However, the policy analysis in chapter 3 indicates that environmental safety appears to be a cross-cutting issue in both domains for two main reasons, which has not been discussed by previous researches yet.

Firstly, nuclear energy-related environmental issues pose considerable risks to the enjoyment and realization of human rights despite sophisticated safety technology. The fact that Japan has witnessed five nuclear accident and incidents since 1981, which have each contaminated the environment (Elliott, 2013), indicates that nuclear energy is hazardous to environmental protection. When nuclear accidents or incidents happen, radiation is released into the atmosphere (Chino et al., 2011) and contaminates all aspects of the environment including air, soil, ocean, underground water resources, and so on, in which people live as evidenced in Fukushima and Chernobyl (Steinhauser et al., 2014). From the EHRs theory perspective when people do not have access to safe water, food, or air due to radiation, then they are being prevented from enjoying basic human rights such as the right to life and health, as ultimately human beings are dependent on the environment to meet basic needs such as food, air, and water for their survival (Anderson and Boyle, 1996; Shelton, 2008). So, nuclear energy poses a risk to the realization of human rights due to accident risks. Accident risk, however, is common to all technologies within industrial societies, which is referred to as a risk society by Ulrich Beck and Anthony Giddens. It might be convincingly argued that all man-made environmental accidents can result in the contamination of nature, which in turn may affect human rights—the Jilin chemical plant explosions in 2005, for example, killed six, injured dozens, and required the evacuation of tens of thousands of residents due to the resultant environmental contamination (Duan et al., 2011). This means that risks to human rights associated with the environment are not limited to nuclear energy alone; any man-made or natural events may affect human rights as nothing has zero risk in the world. The issue is related to the mitigation of risks to achieve environmental safety (Kasperson and Kasperson, 2001). In other words, the main risk to the enjoyment of human rights in connection with the environment is the lack of safety or poor safety which can cause environmental contamination, not nuclear energy itself. For this reason, environmental safety can be seen as a common goal for Japan's energy and HRP.

What constitutes a safe environment, however, may differ from person to person as environmental risks are perceived subjectively (Lundgren and McMakin, 2018). The risk and benefit perception of individuals is directly associated with values influenced by social, political, and economic factors, rather than any form of objective scientific calculation (Xu et al., 2017). However, the main criterion used to determine environmental safety for people in nuclear energy matter *is* based on scientific measures, which is radiation levels. There is no universal standard, but a variety of epidemiology studies show that the average annual dose a person receives is less than 5 mSv/yr (Cuttler and Pollycove, 2009). According to the World Nuclear Association (2015), natural background levels are mostly in the range 2–3 mSv/yr. It is estimated that a level between 5 and 20 mSv may adversely affect human health and result in serious illnesses, particularly cancer (Akyuz, 2017). The Japanese government has established a radiation dose limit of 20 mSv of radiation a year, which is curiously not in line with international standards (Boice Jr, 2012). Numbers regarding radiation levels, however, may be not the main criterion by which to determine how safe the environment is for individuals because people may perceive even low radiation levels as being unsafe. There are many factors which influence Fukushima residents' perceptions of environmental safety and risk including contamination of the food and water, increase in mental and physical health issues among people in the THUs, ongoing radioactive contamination in the exclusion zone, uncertainty about the decontamination process, rumors, propaganda by antinuclear groups about the accident in particular on social media, and so on (Murakami et al., 2015). Therefore, it shows that the above public perception of environmental safety and risks can be different from the safety standards set and regulated by the Japanese government.

Secondly, it can be argued that nuclear energy contributes to environmental protection and the enjoyment of human rights due to its contribution to the reduction of greenhouse gas emissions, which has certain impacts on the realization of human rights when environmental safety is properly implemented. As explained in chapter 3, Japan's NEP is based on the 3Es plus S policy. Some scholars argue that nuclear energy is an instrument to combat climate change issues to achieve environmental protection (Kingston, 2013). There is some empirical research that climate change affects human rights worldwide (Humphreys and Robinson, 2010). Nuclear energy is key to greenhouse gas emissions (Lenzen, 2008), meaning that it contributes to the mitigation of climate change. If nuclear energy is not used, greenhouse gas emissions might increase (Socolow and Glaser, 2009) which in turn may affect human rights, as evidenced by previous studies (Hassan and Khan, 2013).

So, a refusal to adopt nuclear energy may also be a risk to the protection of human rights in Japan. This, however, does not imply that nuclear energy is the only source for the mitigation of CO_2—critics, for example, argue that there is no need for nuclear energy to mitigate CO_2 (McGowan, 1991) because this can be achieved by renewable energy, which also does not release CO_2 (Quaschning, 2009; Bevan, 2012). It, however, does not seem realistic to expect Japan to shut down all NPPs in the short term as they meet 30% of total electricity demand (Chanlett-Avery, 2010), as otherwise, the country may face an energy shortage, which in turn would affect economic growth, national security, and living conditions significantly (Kingston, 2013). Nuclear energy, therefore, can contribute to environmental safety, which is essential for the enjoyment of human rights, in terms of CO_2 reduction while posing risks to the environment due to the potential for nuclear accidents, the issue of nuclear waste, and so on (McCombie and Jefferson, 2016).

Human Rights Violations Raised by the FNA

As discussed above, environmental safety is the common concern of Japan's HRP and NEP. My empirical research shows that this is indeed the case for the FNA matter. This research has found empirical evidence that the lack of sufficient environmental safety associated with the FNA has affected the enjoyment of human rights including the rights to life, health, and property.

The literature regarding the FNA is broadly divided into three categories. The first of these investigated whether the accident resulted in environmental contamination through the release of radioactive materials into the ocean, air (Steinhauser et al., 2015), soil (Yasunari et al., 2011), and groundwater (Shozugawa et al., 2016). The above research confirms that the accident contaminated the natural environment. The second category investigates the impact of the accident on human health. There has been considerable research on the theme that the accident has caused an increase in cancer, mental health issues, diabetes, hypertension, obesity, and so on (Hasegawa et al., 2015; Maeda and Oe, 2017). Third, there are a few research efforts that have focused on how the contamination of the environment has affected the nonhuman world (Mousseau and Møller, 2014). They confirm that it has resulted in an increase in the mortality of birds, butterflies, and various insect species. To the best of my knowledge, there is no empirical research that has explored how the FNA affected the enjoyment of SHRs, including the rights to life and health. There is also very little research that has addressed the possibility as to whether the accident may have caused human rights violations in the region. Khan (2018), for example, argues that people have a right to freedom of movement but that the government forced them to evacuate, which is a violation of the right to freedom. He did not note any linkage between human rights violations and the contaminated environment in Fukushima.

Human rights violations in connection with the accident, however, have been pointed out by civil society organizations and the UN. For example, the UN Human Rights Council, which observed the situation relating to human rights of Fukushima nuclear victims in 2013, confirmed that victims are not able to realize the right to the highest attainable standard of physical and mental health guaranteed by the ICESCR due to the radiation dose limits, which can be up to 20 mSv, and poor access to accurate information about radiation and its effects on human health (OHCHR, 2013). Similarly, Human Rights Now, a Tokyo-based IHR NGO, submitted a written statement on Fukushima for the upcoming fortieth session of the UN Human Rights Council entitled "People Affected by the Fukushima Nuclear Disaster Continue to be Forced to Live in Unsafe Areas" that noted the right to adequate housing and the right to health have been affected by the ongoing radioactive contamination in the region (HRN, 2019). These organization's claims are in line with my finding that radioactive contamination is the principal reason behind human rights violations in the region. In a similar manner to this book, they found that the ongoing contamination has affected people's right to health in Fukushima. However, none of these organizations has provided any comprehensive academic research into the human rights violations actually caused by the FNA; rather, they generally only focus on the right to health. However, this book has evidence that ongoing contamination also affects the rights to life and health in the region.

Recognition of the linkage between human rights and the environment may bring both advantages and disadvantages. Firstly, an increasing emphasis has been made by scholars that human rights are a widely accepted and well-established set of legal norms, which has strong connections to mechanisms of implementation and enforcement because human rights law holds instrumental value as a source of positive law and is backed by a series of adjudicative mechanisms allowing individuals and civil organizations to claim their rights guaranteed by the law at national levels (Anton and Shelton, 2011; Atapattu, 2002; Bosselmann, 2001; Woods, 2006). In other words, the advantage of the human rights framework derives from the fact that it creates a rights-based approach (Woods, 2006) because the legal system of human rights includes a broad spectrum of obligations that are binding upon states both "positively" and "negatively" (Pathak, 2014). This seems the case in Japan because as indicated in the policy analysis in chapter 3 human rights are well established and guaranteed by Constitution of Japan. Article 11 of the Constitution of Japan states that "*the people shall not be prevented from enjoying any of the fundamental human rights*" (KANTEI, 2018), which means that the state is responsible for ensuring that third parties do not interfere with their enjoyment, as discussed in other research (Anderson and Boyle, 1996). According to such research, there are many examples whereby the Japanese courts have found that the Japanese government has been liable for violations of human rights, and that public institutions should take sufficient measures to protect the human rights guaranteed by the constitution. For example, after the Japanese government's exclusion of the Osaka Korean High School from its high school education tuition waiver program, the Osaka court found that this violated the right to the education guaranteed by the constitution (Tai, 2007). Accordingly, establishing a link between human rights and the environment may create a rights-based approach that the developed compliance mechanisms permit individuals and groups to claim their human rights affected by the accident (Anderson and Boyle, 1996). Many researchers note that the Japanese government has clearly failed to protect human rights and residents have not been able to enjoy their rights in this regard such as the rights to life and property (Omoto, 2013; Funabashi and Kitazawa, 2012; Aldrich, 2013). It is also legally recognized, via the Japanese court itself, that both TEPCO and the government are guilty of negligence in contributing to the extent of the Fukushima nuclear disaster, and has ordered them to pay damages (Acton and Hibbs, 2012). Therefore, this gives opportunities to the victims to sue the government, or TEPCO, for human rights issues because the Japanese state has legal duties to protect human rights from the effects of harmful activities including nuclear accidents within their jurisdiction.

Secondly, framing the nuclear accident as a human right violation can be used by particularly concerned citizens to compel public institutions to improve regulations regarding environmental contamination and risk, secure higher standards of environmental quality, and to take measures to control contamination affecting people (Cullet, 1995; Glazebrook, 2009; Woods, 2010) because public institutions have positive obligations to protect, promote, and fulfill human rights, as well as negative obligations to abstain from rights violations (Anton and Shelton, 2011). The policy analysis in chapter 3 shows that the Japanese government has the duty to guarantee the creation of the conditions that may be necessary in order to prevent violations of an inalienable right such as the right to life and health. On this basis, as the nuclear accident poses barriers to realizing human rights, then victims of a contaminated environment can force their government to secure higher standards of environmental quality; to take measures to control contamination affecting people, as governments become directly accountable for their failure to regulate and control environmental nuisances, including those caused by corporations; and for facilitating access to justice and enforcing laws. For example, the empirical findings in chapter 4 show that residents believe that they live in an unsafe environment, which would nominally appear to be an obstacle to the enjoyment of the right to life, and which is in line with previous research. According to Sato and Lyamzina (2018) and Sugimoto et al. (2014), the majority of residents do not find the environment around the THUs to be safe due to the prevalent levels of radiation, despite the fact that the Japanese government declared that it is safe. THUs are a state policy, which makes the government responsible for the human rights problems that residents face (and have faced) in relation to the contaminated environment (Kingston, 2013). In other words, the nature of the environment around the THUs also restricts their lives in Fukushima. Thus, establishing a link between human rights and the environment in the FNA matter may force states to preserve the physical environment upon which human rights depend as public institutions have positive obligations to protect, promote, and fulfill human rights.

Thirdly, the TA presented in chapters 4 and 5 indicates that residents do not have any awareness about the human rights violations in connection with the contaminated environment in Fukushima. Similarly, other research indicates that residents only tend to apply to courts with regards to compensation issues, rather than the violation of human rights in connection with the FNA. Kingston (2013), for example, states that people tend to apply to courts for compensation for the evacuation and restriction orders on their properties. By the same token, some empirical research shows that the main issues raised by residents include (1) poor living conditions in the THUs, (2) poor compensation, and (3) no safe access to food and water (Feldman, 2012; Osaka, 2012). This means that human rights issues are not on the residents' agenda. However, this awareness exists among NGOs. My interviews with such organizations show that human rights NGOs, such as Greenpeace and Human Rights Watch, have filed lawsuits to sue the government for human rights violations raised by the FNA. Taking a human rights approach to the Fukushima accident might be advantageous in that it may increase public awareness of human rights issues in connection with the environment (not just radioactive contamination) in Japan (Woods, 2006). The country has serious environmental problems such as air pollution and climate change like other industrialized countries. The fact that nuclear accident has caused human rights violations in connection with the environment may increase public awareness of human rights issues in connection with other environmental issues such as climate change and global warming. Additionally, if people become aware about human rights violations in connection with the FNA, individuals may apply to the courts to seek appropriate remedy for human rights issues and to encourage public institutions to ensure environmental safety (Anton and Shelton, 2011; Hajjar Leib, 2017). For example, if a court decides that people do not enjoy the right to health in the THUs due to the quality of the environment, this may force the government to build new THUs in a safer location. However, this does not mean that human rights are the only way to force the government to build new THUs, namely, through the medium of legal action. Japan has appropriate environmental law, and indeed a large number of people sued the Japanese government for the impact of different environmental problems on their health through environmental law. Clearly, residents may find legal remedies within environmental law; the main difference, however, is that human rights law has stronger implications because these rights are guaranteed by the constitution, which is the highest law in Japan, and by legally binding IHRLs that Japan must implement.

Fourthly, the human rights approach to the FNA may provide a solution to the distributive (in)justice faced by the Fukushima residents. The TA showed that while the FNA has impacts on nature, a contaminated environment is certainly not shared by individuals equally, which raises a distributive justice issue (Feldman, 2012). My analysis shows that the radiation levels in the THUs are higher than the normal background levels, and the living conditions themselves are quite poor, which affects their lives and health, in addition to being denied access to their property in the exclusion zone due to the environmental impact of the FNA. In contrast, those with enough economic means live outside the contaminated area in places such as in Tokyo, which is relatively much less contaminated (Kingston, 2013). All citizens benefited from electricity generated by the Fukushima Power Stations equally, regardless of their location, wealth, and so on, but the harm of the FNA is not equally distributed across Japan (Harada et al., 2014). This results in distributive (in)justice, which is the idea of balancing benefits (e.g., clean air, green spaces) and costs (e.g., pollution) associated with the way in which we disseminate and consume environmental entitlements (Dobson, 1998; Cotton, 2017). Very little research has given appropriate attention to this topic, namely, that the unequal distribution of environmental burdens throughout society is a dominant issue in the FNA matter. Kingston (2012), for example, showed that there are certain groups in Fukushima that are disproportionately exposed to, and affected by, radioactive contamination, which in turn affects their living conditions. The main principle of human rights is, however, that every person is free and equal in dignity and rights; they are entitlements of all people, regardless of their wealth and social status. Article 14 of the Constitution of Japan, for example, states that "*All of the people are equal under the law and there shall be no discrimination in political, economic or social relations because of race, creed, sex, social status or family origin*" (KANTEI, 2018). That is, Article 14 enables even "marginalized" and "disadvantaged" social groups to claim and enjoy the rights that might otherwise have been affected by the accident. Human rights, therefore, can be equally applied by any social group to address the unequal distribution of nuclear harm to their lives, health, and property, as guaranteed by Constitution of Japan.

Fifthly, another important point is related to the fact that human rights may reduce the problem of moral relativism over environmental contamination raised by the FNA. The ethical issues over nuclear energy mainly focus on the comparison between the benefits and risks of NPPs, the catastrophic impact of nuclear accidents, disclosure of facts, and safety regulations, which are substantially moral judgments (Schreurs and Miranda, 2014). There are, however, different moral norms over NPPs and the environment as to what is acceptable and not acceptable, or right and wrong, about nuclear energy and the environment being dependent on the value of the people (Kermisch and Taeb, 2017). Indeed, my analysis indicates that officials perceive environmental risks associated with the FNA differently from residents. That is, nuclear energy-related environmental issues are, ethically, a controversial topic. Similarly, human rights are also moral principles or norms that describe certain standards of human behavior (Zarsky, 2012). It, however, differs from NPPs and the environment in that there is a common international consensus over the fact that they are expressed in internationally recognized human rights declarations, covenants, conventions, and charters; this means, in turn, that they are universally recognized and accepted (Renteln, 2013). This, however, does not mean that the moral standards presented in human rights are necessarily true, it only means that they are broadly accepted all over the world. When something happens in our societies, it is more fashionable to say that it is wrong because it is a violation of human rights rather than an immoral act (Anton and Shelton, 2011). Therefore, if the FNA is defined as a human rights problem in connection with the environment, then human rights may reduce the problem of moral relativism over environmental contamination raised by the FNA because, ultimately, human rights are broadly and universally accepted, as discussed by Steiner et al. (2008) and Slaughter (2009).

A purely human rights-based approach to the accident may result in a number of discrepancies. Firstly, it can be argued that there is no need for a human rights-based approach to force the government to secure environmental protection because the policy analysis in chapter 3 shows that Japan has a NEL, which obliges the country to take sufficient precautions in order to achieve environmental and nuclear safety, which has been studied previously (Suzuki, 2011). Several research efforts have argued that nuclear safety can be achieved through law on nuclear energy in Japan because there are several hard and soft laws that regulate safety standards and set international benchmarks to which Japan must subscribe (Irie, 2011). Indeed, the Convention on Nuclear Safety, adopted by IAEA in Vienna on June 17, 1994, is, for example, the world's first binding international standard on the siting, construction, and operation of land-based nuclear facilities (Kamminga, 1995). It imposes a set of obligations on Japan to implement certain safety rules at all civil facilities related to the use of civilian NPPs. Similarly, Japan has "the

Basic Environment Law" which provides details about basic national policy concerning the environment (MOE, 2019). These regulations mean that Japan is already obliged to secure nuclear and environmental safety through nuclear and environmental law. Prevention is, however, important to mitigate accident risks and their impact on the environment, but human rights law in Japan does not specify precisely what measures should be taken or what technical requirements are needed for nuclear and environmental safety. This means that a purely human rights approach to the environmental or nuclear safety may not be an effective instrument to mitigate risks associated with nuclear energy and the environment. Accordingly, it can be argued that there is no need for human rights to achieve nuclear energy safety because the law on nuclear energy in Japan already regulates this.

Another challenge to the human rights approach, as Glazebrook (2009) and Shelton (2008) argue, is that there is some considerable difficulty with the issue of causation and proof as to how environmental contamination affects the realization of human rights. In other words, it is hard to prove the extent to which environmental contamination has contributed to the violation of human rights. Indeed, the TA in chapter 4 indicated that this is the case in the case of the Fukushima accident because residents stated they have physical health problems due to the accident but there is no mounting scientific evidence regarding the environmental threats to their physical health. Fukuda and Fukuda (2012), for example, note that there may be many reasons which affect physical health including stress, unhealthy food consumption, smoking, poor sleeping habits, genetics, and alcohol consumption. Similarly, Tateishi (2015) argues that people live with the fear of radiation, but the 2011 tsunami and earthquake event, or other potential accident risks, may trigger fear among residents. In other words, radioactive contamination may not be the only reason behind human rights violations as other factors, such as tsunami, earthquake, displacement, and so on, may also affect people's health and lives. As discussed by other empirical studies (Sato and Lyamzina, 2018; Sugimoto et al., 2014), the ongoing radioactive contamination may be the main reason behind health issues such as depression and anxiety, but the questions as to how this could be reasonably proven and to what extent the contaminated environment has contributed to these health problems remain unanswered.

On the other hand, the discussion above does not necessarily mean that all people in Fukushima or in Japan have suffered from human rights violations in connection with the contaminated environment. The impact of the radioactive contamination on people's rights to life, health, and property may vary from location to location or person to person because the associated radiation levels differ by region in Japan, as indicated by Harada et al. (2014). What is more certain is that radioactive materials were released into the environment from Fukushima because of the radioactive cloud that spread over the whole country and, indeed, the world. Ishida and Yamazaki's research (2017), for example, shows that radiation levels in Tokyo have increased since the Fukushima accident. However, there is no research or real evidence that those who live outside the Fukushima region have suffered human rights violations in connection with the accident. This means that the contaminated environment-related human rights issues may be limited purely to the THUs or neighborhoods near the power stations, or that there is a further need to investigate whether the FNA has affected people's human rights in different cities of Japan.

Secondly, the above discussion does not suggest that all residents have suffered to the same extent from human rights violations. The degree of violation of human rights that residents have felt or faced is actually somewhat nuanced. For example, some empirical research shows that those who have children suffer from higher levels of psychological distress than those who do not because mothers are concerned about not only about themselves but also their children's lives and health, as thyroid cancer is common among children in Fukushima (Goto et al., 2014; Sugimoto et al., 2014). This is also the case for the violation of the right to property. My research in chapter 4 shows that while some people have both luxurious houses and working places in Fukushima, others have only one small flat, each of which clearly has different economic values. This means that those who have more valuable property may be affected to a greater extent than others in terms of the violation of the right to property. Similarly, those who have children suffer from mental health issues such as depression to a greater extent than those who do not have children because, as my research and similar studies show, mothers are understandably very concerned about their children's health due to the fear of radioactive contamination. All this means that the degree of violation of human rights raised by the FNA can differ from person to person.

REINTERPRETATION OF HUMAN RIGHTS VERSUS RIGHT TO THE ENVIRONMENT

The above discussion shows that rights to life, health, and property have been affected by the contaminated environment. This means that human rights and quality of the environment are intimately connected to each other with regards to the FNA. The environment appears to be an essential instrument for the protection of people' lives, health, and property because the FNA has resulted in contaminated air, soil, and water resources that people otherwise need to meet their basic survival needs (Steinhauser et al., 2015; Yasunari et al., 2011). In other words, the poor quality of nature, as a result of the

FNA, has affected and posed risks to people's lives, health, and property in Fukushima. Some scholars argue that the connection between environment and human rights leads to the specific articulation of human rights in such a way that all mankind has a right to live in an environment that is adequate to ensure their health and well-being as people cannot exist without the nature (Anton and Shelton, 2011), as discussed in chapter 2. In other words, it is argued that a new substantive right to the environment should be added to the list of internationally recognized human rights, such as the right to life, liberty, education, and so on (Anderson and Boyle, 1996). The right to the environment has been recognized by fifty-six countries and various regional documents such as the African Charter as human rights to date (Boyd, 2011). The main rationale behind human rights theory is to protect the fundamental freedoms and dignity of both individuals and groups. This book shows that people's lives and freedoms are at risk due to the lack of a safe environment in Fukushima which can expand the scope of human rights in a way that requires people to have a right to a safe environment, as people's freedoms cannot be enjoyed or realized in its absence.

Some scholars argue that there is no need for a distinct right to environment because a safe environment is already embodied in internationally recognized human rights, such as the right to life, as discussed in chapter 2 (Boyle and Anderson, 1996). This discussion is characterized by theoretical, rather than empirical, observations (Boyd, 2011). My empirical data contributes to the discussion that right to a safe environment is an extension of the rights to life, health, and property in Japan because they have been restricted, affected, and, indeed, violated by the ongoing environmental contamination or the lack of a safe environment in Fukushima. We found evidence that the quality of the environment raised by the FNA has affected people's lives, health, and property, or in other words three of the fundamental human rights that are guaranteed by Constitution of Japan and IHR document, such as the ICCPRs. This means that internationally recognized human rights, including the right to life, health, and property, already require that a safe environment be enjoyed and realized by individuals, and the contaminated environment has affected their realization. This can be interpreted in such a way that if people have a right to life or health then they also have a right to a safe environment because these human rights can only be fully enjoyed in a sound environment. While Boyd (2011) refers to this as greening existing human rights, Anderson and Boyle (1996) refer to it as a reinterpretation of human rights in an environmental context. Such interpretations of established human rights, effectively through the medium of the environment, can be also observed in regional and international human rights institutions such as the European Court of Human Rights, with regards to which the European

Court has ruled that enjoyment of Article 8, which guarantees that private and family life requires a safe environment (Caglayan, 2015).

However, this is criticized by some scholars in the sense that a reinterpretation of human rights may remain ineffectual in terms of addressing environmental contamination which does not affect human rights or human beings (Boyd, 2011). Indeed, the FNA clearly affected the nonhuman world, and nature does not pose a direct risk to people's rights to life, health, and property, as underlined by various research efforts (Aliyu et al., 2014). This remains beyond the scope of human rights and indeed appears to be a limitation to the reinterpretation of human rights to address environmental safety. This, however, cannot be addressed by the right to the environment either because these rights indicate that people have a right to live in a clean environment that is favorable to their development. In other words, both a reinterpretation of human rights and a distinct right to the environment are anthropocentric, seeing the environment as a tool to meet human needs rather than requiring the protection of the environment for its own stake (Woods, 2006). In other words, neither a reinterpretation of human rights nor a distinct right to the environment can effectively address the environmental impact of the accident on the nonhuman world.

This does not suggest that a distinct right to environment is useless. Many countries simultaneously recognize both a distinct right to environment and other human rights, such as the right to life and health, in their constitutions, such as Turkey and South Africa. As Boyd (2011) noted, a new right to the environment can only complement, enrich, and enhance preexisting rights, but it does not bring any new insight into the relationship between human rights and the environment because existing human rights already require a safe environment, as evidenced by this TA; this may oblige countries to achieve environmental safety by reducing radiation levels in certain regions, for example, or construction of THUs which have lower radiation levels than required by international standards because states have a positive duty to protect human rights. The right to environment is, therefore, already embodied in the right to life, health, and property, and which have been affected by the lack of a safe environment in Fukushima.

Therefore, this research has for the first time found that the FNA has caused human rights violations in connection with the environment. Governments have a positive obligation to protect human rights, meaning that states can be forced to achieve a safe environment in order to protect human rights because human rights require a safe environment to be enacted, as shown in chapter 4. Universal human rights, including the rights to life, health, and property, are well established in Constitution of Japan. According to this study, the realization of these human rights, then, is dependent on the quality of the environment in Fukushima. In other words, Japan needs to achieve a safe environment in order to protect existing human rights as they require a safe environment. My finding shows that the country does not need a distinct right to the environment because a safe environment is already essential to the realization of existing human rights in Japan as evidenced in Fukushima.

Main Finding 2: The Linkage between Substantive Rights and PRs

This research found that PRs have been violated by both the government and TEPCO to some degree, which has previously been studied by a number of scholars, as discussed in chapter 5. There has been some research into PRs and substantive rights in the Fukushima matter, but they do not consider the linkage between substantive rights and PRs or the human rights aspect of these linkages in any way. What this research has found for the first time is that the violation of PRs has triggered an associated violation of SHRs. The following section consists of two main parts. It first argues how the violation of the right to information has similarly contributed to the violation of the right to health. It then discusses how the lack of public participation in associated decision-making processes has led to the violation of the right to health in Fukushima. By doing so, we critically assess the contribution to the literature that the enjoyment of SHRs is dependent on the enjoyment of PRs in Fukushima, which means that substantive rights and PRs are not alternatives to each other, and can be interrelated. This contributes to answering the main research question, "What are the advantages and disadvantages of framing nuclear energy policy and its possible impacts, in the context of environmental human rights?," in the sense that PRs are essential to the enjoyment of SHRs in the nuclear energy governance of Japan.

The Linkages between Violation of the Right to Information and the Enjoyment of the Right to Health

Chapter 5 shows that both the government and TEPCO have violated the right to information in two ways. Firstly, they did not share emergency information with the public in the first few days of the accident, which is in line with the extended literature regarding the FNA (Funabashi and Kitazawa, 2012). Yilmaz (2011) and Pierpoint (2011), for example, state that TEPCO withheld information from the public; the public authorities did not communicate with the public in an accurate manner to inform them about the risks associated with the accident; and members of the public who lived within the 30-km circle around the site of the accident were not informed about the need to evacuate due to the lack of any advanced preparation for emergency. People who lived near the damaged stations were not informed about the risks associated with the NPPs and what they should do or should not do in the instance of a nuclear accident (Figueroa, 2013; Khan, 2018), which is referred to as an "information disaster" by Hatanaka et al. (2014). The accident happened at 2:46 p.m on March 11, 2011, but the evacuation of residents within a 3 km radius of the accident was only ordered at 7:45 a.m on March 12 (Hatanaka et al., 2014). This area has been declared a "difficult to return zone" because the annual integrated doses are over 50 mSv, well in excess of any recognized national or international standards. Accordingly, residents were unnecessarily exposed to radiation as they were not informed about the need to evacuate on the same day as the accident. In other words, residents continued to live in an unsafe environment due to the lack of any timely sharing of information. This elevates their risk of developing cancer, as stated by Murakami et al. (2015) and Shimada et al. (2018), but there has been, however, no real certainty about how much radiation people received in the first days after the accident and how this radiation has affected people's mental and physical health. What is more certain is that being exposed to radiation for longer periods increases the chance of subsequently developing health problems. States have a positive obligation to supply the essential resources that allow individuals to enjoy their inalienable human rights (Camur, 2017). This, of course, includes the disclosure of information in the instance of a nuclear accident because residents will clearly not have the opportunity to measure radiation and its risks to their health. For this reason, the lack of timely sharing of information resulted in people being unnecessarily exposed to the radiation that increased their risk of cancer-related health issues. This fact was also underlined by the UN Human Rights Council which observed the situation for human rights of Fukushima nuclear victims in 2013. The Special Rapporteur confirmed that victims have not realized their right to the highest attainable standard of physical and mental health otherwise guaranteed by the ICESCR as the government and TEPCO have not to date shared detailed information about the risk of radiation to people's health which "*may give children and parents a false sense of security*" and/or "*may result in children's exposure to high levels of radiation*" (OHCHR, 2013, p. 17). Similarly, the Human Rights Committee (CCPR), established under the International Covenant on Civil and Political Rights, recommended, "*The State party should monitor the levels of radiation and disclose that information to the people affected in a timely manner*" (OHCHR, 2014). Therefore, timely sharing of information in an emergency situation is essential to prevent the violation of the right to health, but which did not occur in Fukushima.

Violation of access to information is not limited to the timely sharing of information. The literature and IEL documents such as the Aarhus Convention indicate that information should be scientifically accurate and provable (Mason, 2010). If a public institution shares nonscientific information with the public, this can in itself be regarded as a violation of the access to information (Cramer, 2009). In other words, the reliability and accuracy of the information are as important as the public disclosure of the information itself (Mason, 2010). My research, and indeed similar studies, show that this is the case in the Fukushima matter because the Japanese government determined that radiation levels between 1 and 20 mSv are safe for human health, despite the fact that this is not consistent with international standards or other countries' regulations regarding nuclear energy, such as the United Kingdom and the United States (Sutou, 2016). There has been no scientific research into whether a radiation dose between 1 and 20 mSv is safe for human health (Kingston, 2012). This issue was also underlined by the Special Rapporteur of the UN Human Rights Council, as discussed above, due to the radiation dose limits being up to 20 mSv, which is not consistent with international standards or epidemiological research (OHCHR, 2014). The THUs were constructed in neighborhoods which show radiation doses of less than 20 mSv according to the Japanese government. However, this research and others (Kingston, 2012) have found that people do not consider the environment in which they live to be safe because a "safe" radiation level of up to 20 mSv has no empirical scientific backing. Accordingly, the lack of scientific information regarding radiation safety levels has triggered mental health problems because residents think that they are being exposed to excessive radiation in the THUs, which has consequently caused depression, anxiety, stress, and so on, among residents. This has previously been studied by various studies. Suzuki et al. (2015), for example, show that psychological distress is quite common among residents because they fear being exposed to the radiation they perceive as being present in the THUs, and thus do not consider their environment safe. THUs were constructed in neighborhoods that were declared safe by the government but which residents think are unsafe, which has resulted in the understandable fear among the public of being exposed to excessive levels of radiation. Therefore, the associated lack of disclosure of scientifically proven information has become an obstacle to the enjoyment of the right to health in Fukushima, regardless of whether the 20 mSv level is, in fact, safe.

The linkage between the enjoyment of SHRs and access to information has been recognized by previous IHR courts' decisions. In the case of *Yanomami v. Brazil*, the Inter-American Commission on Human Rights to the Court, for example, ruled that the Brazilian government violated the right to life under Article 1 of the American Declaration by constructing a highway and allowing mining in Yanomami territory. It concluded that human health requires that individuals have access to information regarding environmental impacts of projects which are likely to affect them (Albert, 1992). Similarly, in the case of the *Oneryildiz v. Turkey*, the ECHR indicated that the government violated the right to life under Article 1 of the ECHR because Turkey did not share information regarding the risks and impacts of a methane explosion in

1993 at a municipal rubbish tip in the slum quarter of Istanbul (Caglayan, 2015). These and similar court decisions indicate that the right to information is a fundamental requirement of the realization and enjoyment of SHRs. This book provides similar cases that demonstrate the right to health requires the enjoyment of PRs, including the right to access to information in Fukushima because the violation of such has contributed to the violation of the right to health.

This research reports, for the first time, that the violation of the right to information in Fukushima has affected people's SHRs, particularly the right to health. This ultimately indicates that enjoyment of the right to health is dependent on the enjoyment of access to information. In other words, the right to information is essential to the enjoyment of human rights.

The Lack of the Right to Public Participation and the Enjoyment of the Right to Health

The policy analysis in chapter 3 shows that people do not have any meaningful right to public participation in decision-making processes. Similarly, chapter 5 provides empirical evidence that residents have never participated in decision-making processes in any meaningful manner, as has previously been studied by a number of scholars. While Nakamura (2018) confirms that two deliberative opinion polls were conducted at the national level, White (2015b) concludes that the public does not have any important role in terms of its influence over nuclear energy-related decisions. The literature regarding the FNA shows that there has been no meaningful public engagement in which residents and officials could exchange their perceptions of the risks, concerns, and priorities which can be reflected in nuclear energy-related decisions (Kingston, 2013). However, how the lack of meaningful participation in nuclear energy formulation process has affected the enjoyment of SHRs in Fukushima has not been discussed to any sufficient extent as yet. This research has for the first time identified the linkage between the lack of public engagement, particularly in risk-based decisions and the violation of the right to health in the FNA.

Chapter 4 shows that people suffer from mental health issues because they believe that they do not live in a safe environment, as discussed above. They fear that they are being exposed to radiation. By contrast, officials and TEPCO managers think that there is no serious risk associated with the environment on people's lives. In other words, residents' subjective risk perception differs from the national standards that regulate radiation levels in the country. While officials claim that radiation levels are sufficiently safe in terms of human health, residents think that they do not live in a safe environment which in turn triggers mental health problems that include stress and depression, as explained in chapter 4. This has previously been studied by various research efforts into the FNA. Suzuki et al. (2018), for example, show that residents have different subjective judgments about the characteristics and severity of the risks associated with the FNA to those of officials. Goto et al. (2014) conclude that the perception of the risks associated with radiation was very likely to have caused the immediate effects of stress, depression, anxiety, and so on, among residents. It is clear that there is no consensus as to how safe an environment has to be safe in terms of human health between the public and officials. My research and similar studies have found that this risk perception among residents has increased mental health issues because they think that the environment in which they live is not safe, while officials believe the opposite (Suzuki et al., 2018). This means that the lack of socially acceptable risk triggers mental health problems. The fact that people do not participate in risk-associated decisions has resulted in the lack of socially acceptable risks (Figueroa, 2013), which itself results in the fact that residents do not believe they live in a safe environment. Safety levels are determined solely by central government rather than any form of consensus with the public, which is the reason why residents and officials have different perceptions of environmental safety in Fukushima (Hasegawa et al., 2015).

This, however, does not mean that the environment is unsafe, or people are being exposed to unsafe levels of radiation. As discussed in chapter 4, and as indeed stated by other researchers such as Suzuki et al. (2015), this is only about public perception. Risks and benefits can be perceived differently according to people's values, experiences, and so on (Slovic, 2000). Residents and officials have different risk perceptions of the radiation levels. The most common way of determining socially acceptable risk is to engage the public in the associated decision-making process and exchange their concerns, priorities, and so on, as stated by Rogers (1998). This is also recommended by the Rio Declaration, and which Japan is a party to, whose Article 10 states: "*Environmental issues are best handled with the participation of all concerned citizens, at the relevant level*" (EUROPA, 2018). Public participation is seen as essential for proper environmental governance.

Many scholars agree that public participation in the decision-making process may help the government determine what is a socially acceptable risk is for society (Anton and Shelton, 2011; Gellers and Jeffords, 2018). There is clear evidence that NPPs have both risks and benefits which are perceived in different ways by residents and officials. The socially acceptable level of radiation could be determined via public participation, which in turn could mitigate the fear of radiation contamination which affects the right to life and health in Fukushima, and simultaneously mitigate information reliability issues. Lay people tend to assess risk as subjectively expected risks that are not based on statistical data or former experience (Slovic, 1987). Deliberation and negotiation of all the relevant interests, as achieved through public participation, may result in fairer decisions and policy that reflect common values and norms rather than purely scientific ones (Walker, 2007; Whitton et al., 2017). This may help the government determine a socially acceptable risk (Armeni, 2016). Different stakeholders can integrate new perceptions into the discussion, which convey new information, knowledge, and values (Kiss, 2014; Whitton et al., 2017).

Public participation in risk-based decision making is commonly used to identify, evaluate, and resolve environmental health problems through engaging different groups of people in the decision-making process (Sexton, 2013) because the desired quality of the environment is a value judgment (Boyle and Anderson, 1996). It is a two-way communication to define the problem and determine socially tolerable and acceptable levels of risks which can affect the public (Lynn, 1990; Cotton, 2009). It enables decision makers to evaluate both scientific facts and people's values in order to determine which risks are socially unacceptable and what, if anything, to do about them (Sexton, 2013; Whitton et al., 2017; Cotton, 2009). In other words, public participation in risk-based decision making includes two components: (1) the best available scientific evidence and technical judgments; and (2) the concerns, the knowledge, values, and preferences of interested and affected people and parties. When governments, economic players, scientists, and civil society organizations all participate in decisions, these are more likely to be fair and morally acceptable (Renn, 1998).

On the other hand, critics argue that public participation in risk-based decisions may face certain challenges including concern about issues of practicality, transaction costs, and efficient decision making. May (2013), for example, argues that ill-informed, or uninformed, citizens may produce ineffective results in technical topics related to the environment. The fact that people have access to information does not mean that they are well informed regarding environmental events. Nuclear energy is, indeed, a complex technical topic and scientific issue which has both advantages and disadvantages (Akyuz, 2017). Being informed about the benefits and risks associated with nuclear energy seems important to be able to contribute to the energy-making process. Those who participate may not have enough understanding of the issues at stake. Whether people understand the information offered is clearly every bit as important as the disclosure of the information itself (Pierpoint, 2011). However, my analysis in chapter 5 shows that people have both passive and active access to information and only a few people (generally elderly people) find some of this information to be overly complex and technical. Additionally, no research on the FNA has reported that Fukushima residents are uninformed about the FNA and nuclear energy in Japan; furthermore, one of the main aims of public participation is to educate the public

(Armeni, 2016). Public participation can act as a means to inform people about different aspects of a project or decision related to the environment because informed people are also more likely to be able to weigh the environmental pros and cons of these decisions or projects (May, 2013).

Another criticism related to public participation is that the public may be emotionally involved in the decision or issues to be solved (McKinney and Harmon, 2002). This seems to be particularly the case when the projects or decisions have an ongoing impact on people's lives and properties (Gellers and Jeffords, 2018). Indeed, the analysis indicated that nearly all residents raised their objections to nuclear energy and they have a fear of the FNA and radioactive contamination in the region. There are some empirical research findings that some residents do not have a sufficiently objective standpoint from which to make decisions on the density of nuclear energy in Japan as the radioactive contamination raised by the FNA still affects their living conditions (Elliott, 2013). In other words, this research states that people remain angry, and sufficiently emotionally involved, at the Japanese government and TEPCO due to their energy policy to be unable to offer objective opinions. This may raise the issue that people in such a position may, or even will, evaluate any decision related to nuclear energy in Japan from a purely emotional standpoint. However, there are three main points here. Firstly, residents do not have to be objective to participate in the decision-making process because public engagement is about exchanging different values, concerns, opinions, and so on (Lynn, 1990). Secondly, there is no guarantee that officials or TEPCO managers are being, or have been objective. There has been considerable research to indicate that the nuclear village, which includes government officials and the nuclear industry, make decisions that favor the use of nuclear energy in the country, which at the very least can be viewed as partisan (Kingston, 2013; Cotton, 2015). Thirdly, as Hindmarsh (2013) stated, Fukushima is not the only power station in Japan. The country has fifty-four NPPs which are located in different parts of Japan. This means that people near NPPs out of Fukushima may not be emotional about NPPs as they are not affected by the FNA seriously.

Thirdly, it is argued that public participation in each single decision can be time consuming (Lynn, 1990). Public participation requires time because people need to read information related to the plans, decisions, and so on, and participate in the decision-making process to exchange information, values, concerns, and so on (Gellers and Jeffords, 2018). My analysis in chapter 5 shows that residents who work do not have time to participate in the decision-making process or they are not willing to spend time for public participation because overworking is a cultural value in Japan that many people do not have enough free time for other activities. There has been some research that suggests that overwork is has been ingrained as a part of life for the majority of Japanese people (North, 2011; Yamauchi, 2017). Many people may not even take a day off, or may work during their holidays. There has been no formal research into this issue, but it may be argued that these people may not be willing to participate in any decision-making process as they may prefer to spend this time pursuing their careers or work. We, however, found that the THUs house a large proportion of the older population who are typically retired or are housewives who do not work. Similarly, some empirical research suggests that 60% of the THUs house elderly residents (Shinohara et al., 2013). They may represent their families in the decision-making process. In other words, the culture of overworking may not be seen as a common issue in the case of public participation.

Lastly, some scholars argue that the stakeholders who come to the table of discussion may hold very different perceptions of the problem as they may have different priorities and concerns about nuclear energy (Mason, 2012; McKinney and Harmon, 2002). For this reason, it will be difficult to reach a consensus due to the incommensurability of different perceptions and positions (Palerm, 1999). Indeed, the analysis in chapter 5 shows that public demand is different from what the government wants. While residents request nuclear energy-free country, the Japanese government tends to continue with the 3Es plus S policy through the balance between nuclear energy and renewable energy. Their perceptions of the risks associated with the FNA are also different—while residents do not feel the environment to be safe the relevant officials do, as has been studied by previous research regarding the FNA (Hindmarsh, 2013). Aldrich (2013), for example, stated there has been no consensus between the public and pronuclear party over NPPs. The public is opposed to the government's energy policy as they now demand a nuclear-free energy policy.

The fact that enjoyment of SHRs requires the enjoyment of public participation in decision-making processes has been recognized by various national, regional, and international court decisions. In the cases of *Taskin et al. v. Turkey* and *Giacomelli v. Italy*, for example, the European Court of Human Rights held that state parties had violated the right to privacy and family life because their interests or comments have not been given sufficient weight in the decision-making process (EUROPA, 2016). The courts ruled that participation in the decision-making process by those likely to be affected by environmental issues is essential to proper compliance with the SHRs guaranteed by the ECHR (Boyle, 2012). Similarly, in the case of *Lansman v. Finland*, the Human Rights Committee, which is the body of independent experts that monitors the implementation of the ICCPR by its state parties, ruled that Finland had violated the right to enjoy culture under Article 27 of the ICCPR because it failed to ensure effective participation of the affected groups in the decision-making process related to mining and its environmental impacts. The Human Rights Committee's decision shows that SHRs guaranteed by the ICCPR require the protection of PRs. Japan itself has also ratified the ICCPR, which guarantees the right to physical and mental health. The country has a positive obligation to protect human rights guaranteed by the ICCPR. However, my findings show that people do not enjoy the right to health because they believe the environment in which they live is not safe in Fukushima, while officials claim the opposite. Members of the public who are most likely to be affected by risk-based decisions are not involved in the determination of what constitutes the socially acceptable risks associated with the radioactive contamination, which triggers mental health problems.

The above analysis shows that these three PRs can be considered human rights for two main reasons. Firstly, three PRs are already recognized as human rights by the IHR documents that Japan has previously ratified, such as the ICCPRs (Anton and Shelton, 2011). This means that all Japanese citizens already have the human rights to public participation, access to information and justice, as guaranteed by the legally binding IHRL. However, none of these human rights are specified in terms of environmental issues. Indeed, this is not unique to PRs because all human rights do not have any scope in their terms as to where, when, and how they can be invoked by the people. For example, people have the right to expression guaranteed as a human right by various human rights documents such as the UDHR and the ICCPRs (Piovesan, 2004). However, none of the IHR documents specify as to when, how, and where people can enjoy the right to expression because the individual is inherently entitled to their human rights regardless of location, time, and topic. Accordingly, people can enjoy human rights, including the three PHRs, in relation to any given matter, including the environment, education, health care, politics, arts, sports, and so on, which are likely to affect them. Secondly, the above analysis indicates that SHRs require PRs to be realized by individuals. If a person has the right to life, he also needs the right to access to information because a lack of, or poor access to, such information can violate the right to health, as evidenced in Fukushima. This has also been underlined by the courts' decisions. For example, in the case of *Guerra and Others v. Italy*, the European Court ruled that a particular chemical factory had violated the right to private life because it did not provide essential information regarding the risks associated with their chemicals to the public who lived near the factory (ECHR, 2018). Similarly, in *Tătar v. Romania*, the European Court ruled that the government had violated right to the private life because it had not guaranteed its citizens the right to participate in the decision-making processes concerning environmental issues which might have affected them (ECHR, 2019). The above indicates that PHRs are seen as a requirement for the realization of SHRs. This is also supported by my evidence that the violation of the right to information and the lack of public participation have contributed to the violation of the right to health in Fukushima.

Therefore, this research has for the first time identified that the violation of access to information and a lack of the right to public participation in decision-making processes have contributed to the violation of SHRs in Fukushima. This means that the enjoyment of SHRs is dependent on the enjoyment of PRs. My research found evidence that substantive rights and PRs compete each other in terms of achieving a safe environment, an issue which has been not previously discussed in the literature. There is a clear need to improve access to information, which should be scientifically provable, reliable, and shared in a timely manner in order to protect the Fukushima residents' human rights guaranteed by the constitution and IHL ratified by Japan. The government also particularly needs to engage the public in risk-based decision making to determine socially acceptable risk levels which may contribute to the realization of the right to health and life in Fukushima.

On the other hand, the interaction between substantive rights and PRs may not be one-sided as discussed in chapter 2. Some scholars argue that PRs can only be realized through the enjoyment of SHRs (Boyle and Anderson, 1996). In other words, violation of SHRs may also result in the violation of PRs in the case of Fukushima. For example, many people suffer from mental health problems in connection with radioactive contamination, as demonstrated in chapter 4. These people may not enjoy access to information, public participation, or to justice due to health problems, because those who suffer from mental health issues stated that they frequently have to go to hospitals to get assistance, meaning that they may not have enough time to enjoy PRs. Another example is that some residents do not have access to their workplaces in the exclusion zone, which has affected their economic situation as they lost their jobs. Economically marginalized groups may not enjoy an appropriate level of access to information and the courts because they can be charged to access to information and the courts as argued in chapter 2. For this reason, the enjoyment of SHRs may affect the enjoyment of PRs. However, we acknowledge the limitation that this research has only gained direct evidence into only how the violation of PRs has triggered the violation of SHRs. There is a further need to investigate how the violation of SHRs has affected the enjoyment of PRs in relation to the FNA, in particular, the right to information which is guaranteed in Japan. This may help us better understand the connection between PRs and SHRs as pertaining to the FNA matter.

CONCLUSION

This chapter has focused on the main findings of the research by linking the evidence therein to previous research in this field, and how they contribute to further existing knowledge regarding the FNA and EHRs. The chapter reaches two main conclusions. Firstly, this book provides evidence that the FNA is a human rights issue in connection with the environment by answering the main research question, namely, "How does nuclear energy relate to environmental human rights?" The accident has resulted in the violation of the rights to life, health, and property in Fukushima that are otherwise guaranteed by the Constitution of Japan. This shows that a safe environment is essential to the realization of human rights, but which has not been recognized by Japan at the policy and constitution levels. From an EHRs perspective, this means that there is no need for a distinct right to the environment because existing human rights already require a safe environment in Japan, as shown in chapter 4. Human rights are guaranteed by the constitution and IHR documents, such as ICCPR, which impose a certain obligation on the government to protect human rights. In other words, the government is obliged to provide the necessary opportunities which enable people to enjoy those human rights that require a safe environment. This means that as a safe environment is essential to the realization of the rights to life, health, and property in Fukushima, Japan has the obligation to protect the environment that enables individuals to realize their inalienable human rights because the lack of a safe environment results in human rights issues, as evidenced in Fukushima.

We argued that recognition of the relationship between human rights and the environment in the FNA may bring both advantages and disadvantages. Its advantages include: (1) it creates a rights-based approach to the FNA and gives the victims the opportunity to sue the government, or TEPCO, over human rights issues; (2) it enables concerned citizens to compel public institutions to regulate environmental contamination and risks, and secure higher standards of environmental quality; (3) it should increase awareness about human rights issues in connection with the environment in Japan; (4) it may mitigate issues of distributive justice in connection with the FNA; and (5) it could mitigate the ethical discussion over NPPs and its environmental impacts. Its disadvantages include: (1) there may be no need for a human rights-based approach to the environment because nuclear law and environmental law may achieve appropriate safety; (2) there are difficulties with the issue of causation and proof in human rights issues in connection with the environment; (3) the human rights framework does not address nuclear and environmental safety; (4) the human rights approach to the environment is anthropocentric, meaning that the environment is not a priority for its own sake.

Framing the accident as a human rights issue does not mean that a human rights approach to the accident is the only approach to addressing the consequences of the FNA. Environmental law and NEP can address radioactive contamination and safety, while human rights laws can address human rights aspects in connection with the environment. In other words, NEL, environmental law, and human rights law are not alternatives to each other, or somehow mutually exclusive. They are different instruments that can be used by citizens to achieve their common goal, namely, that of a safe environment. In the instance where human rights law does not address the impact of the accident on the nonhuman world, environmental law or other legal instruments can be exercised by concerned people as an alternative. A human rights approach to the FNA, if applied in isolation, may be a somewhat weak instrument with which to solve all the issues connected to the FNA, particularly as the scope of human rights law in Japan does not include nuclear energy safety. Nuclear safety can be achieved by combining nuclear law and environmental law. However, human rights can be effective as an instrument to address energy safety and environmental protection as the absence of these factors may give rise to certain human rights issues.

Secondly, this research has for the first time identified that the violation of PRs has triggered an associated violation of SHRs. Japan has guaranteed the right to information regarding the environment, but the government and TEPCO did not share information in a timely manner in the first few days following the FNA, which caused people to be unnecessarily exposed to the radiation, increasing the subsequent risk of developing radiation-related illnesses such as cancer. In other words, the lack of a timely sharing of information was a violation of the right to health. Additionally, there is no scientific rationale behind the safe radiation levels stated by the Japanese government. This has resulted in the fact that people do not consider their environment to be safe, which has triggered mental health issues as discussed in chapter 4. This means that the lack of disclosure of scientific information represents a violation of the right to health. Similarly, the lack of the right to public participation in the decision-making process has resulted in a lack of socially acceptable levels of risk, which in turn has affected the realization of the right to health in Fukushima because people think that they live in an unsafe environment. In a practical sense, people do not have any right to engage in risk-based decisions; safety levels are determined solely by the central government. As people's risk perceptions are not reflected in the associated decision-making process, residents believe that the officially stated radiation levels are unsafe, which has triggered mental health problems in Fukushima. Therefore, this book provides empirical evidence that the enjoyment of SHRs is dependent on the enjoyment of PRs in the case of Fukushima. Accordingly, Japan's nuclear energy should guarantee the PRs which turn enable individuals to enjoy their SHRs.

Conclusion

INTRODUCTION

This research has considered how the FNA can be approached within the context of EHRs. More specifically, we have analyzed the cross-cutting issue between Japan's HRP and NEP, namely, that of environmental safety; how the FNA has affected human rights, including the rights to life, health, and property; and how people have invoked substantive and procedural EHRs to influence Japan's energy policy and to address the environmental impact of the FNA on their lives. By doing so, we have aimed to advance the EHR's approach to nuclear energy studies (particularly nuclear accidents), which has not been previously studied by research into nuclear energy or EHRs in a comprehensive manner.

The main purpose of this chapter is to summarize the overall findings of the book. This section consists of two main parts. It first attempts to summarize the major findings of this book by discussing how the research questions might be answered and the research objectives achieved. The main points of each of the empirical chapters are highlighted. Second, it considers how this research and its findings contribute to existing knowledge on the EHRs and nuclear energy study from the theoretical, methodological, and empirical perspectives.

SUMMARY OF KEY FINDINGS AND EMPIRICAL CONTRIBUTION

This book has three overarching empirical findings which contribute to the knowledge of human rights issues in connection with the FNA, and further to nuclear energy and human rights policies in Japan. Firstly, we analyzed Japan's human rights and energy policies, both before and after the FNA, which was the first objective this research attempted to address. The first empirical finding which answers the subresearch question in chapter 3, "How does Japan's HRP address EHRs including civil and political rights, PRs, and the right to the environment?" is that Japan guarantees universal human rights, including the rights to life, health, and property, within its constitution. However, the country does not have any policy pertaining to the relationship between human rights and the environment, even though chapter 4 shows that the contaminated environment has affected the enjoyment of a number of basic human rights, such as the right to health, in Fukushima. Neither Japan's HRP nor its constitution addresses the linkage between human rights and the quality of the environment in any way. However, Japan guarantees two PHRs, including the right to information and right to justice, but that public participation in the decision-making process is not currently on the agenda of the pronuclear government.

The second finding, which answers the subresearch question in chapter 3, "How has Japan's NEP evolved since 2011 since the Fukushima nuclear accident?" is that Japan's energy policy has changed considerably since the FNA. These include: (1) safety has become a key principle of Japan's new energy policy, (2) Japan has begun to reduce its dependency on nuclear energy, (3) renewable energy is prioritized, and (4) transparency in nuclear energy governance has now become a key principle of Japan's energy policy.

The third finding, which answers the subresearch question in chapter 3, "How do Japan's NEP and HRP relate to each other, particularly with regards to the matter of the FNA?" is that environmental safety is the cross-cutting issue between Japan's HRP and NEP because nuclear accidents which contaminate the environment affect the enjoyment of human rights, as discussed above. The lack of environmental safety associated with nuclear energy is an obstacle to the enjoyment of human rights. Human rights can be enjoyed or realized by individuals only when Japan achieves environmental safety in the field of nuclear energy, although there is no consensus over how safe is safe enough. This is not, however, addressed by either Japan's HRP or NEP in any way. Human rights and environmental policy integration are recognized at the international, regional, and national levels in environmental issues in particular. This is particularly evident through climate change policy, because there has been certain evidence that climate change affects the enjoyment of human rights worldwide. The Paris Agreement on Climate Change, for example, declares that: "Climate change is a common concern of humankind, Parties should, when taking action to address climate change, respect, promote and consider their respective obligations on human rights."

Secondly, we have identified the fact that there is no previous research that has investigated human rights in connection with the FNA, which is indeed the gap that this research identified. Accordingly, this book has attempted to achieve the research objective of exploring Fukushima residents' experiences of PRs in relation to the Fukushima event by answering two subquestions in chapter 4, including: (1) What are Fukushima residents' experiences of the consequences and ongoing environmental effects of the accident in relation to the rights to life, health, and property? and (2) What are the barriers to, and opportunities for, the realization and enjoyment of human rights in the contaminated areas around Fukushima, particularly with regards to the rights to life, health, and property? The literature regarding the FNA focuses on three aspects of the accident: the associated environmental contamination, health issues in connection with the environment, and the impact of contamination on animals. My empirical research fills a gap in the literature to the extent that the FNA is not only an environmental or public health problem but also human rights violations, in that the contaminated environment has, and still does, affect the enjoyment of the rights to life, health, and property to some degree. In other words, this research provides novel empirical findings that human rights violations in connection with the FNA are evident in Japan.

The TA presented in chapter 4 demonstrates that Fukushima residents do not enjoy their guaranteed human rights due to the ongoing environmental contamination. Constitution of Japan and the IHR that Japan has ratified guarantees a number of human rights including the rights to life, health, work, education, property, and so on. This means that the Japanese government has a distinct obligation to protect human rights. However, the recognition of such is not in itself sufficient to realize human rights as the quality of the environment affects their enjoyment, as argued by EHRs theory. The finding shows that the contaminated environment represents a definite obstacle to the achievement of human rights in Fukushima. People do not, and cannot, enjoy the rights to life, health, and property due to the ongoing radioactive contamination in the region. This helps us understand that the enjoyment of human rights is dependent on the quality of the environment, which is the core argument behind EHRs theory. Radioactive contamination is, however, not the only reason behind human rights violations as other factors, such as poor living conditions, the evacuation process, and tsunami and earthquake in 2011, also affect the enjoyment of the rights to life and health. However, human rights violations in connection with the environment are addressed by neither Japan's HRP nor NEP or the literature regarding the FNA.

Thirdly, this book has attempted to achieve the research objective of exploring Fukushima residents' experiences of PRs in relation to the Fukushima event by answering the research question in chapter 5: "What are the experiences and attitudes of Japanese citizens towards invoking their

PRs in the aftermath of the Fukushima nuclear accident?" We found that residents did not invoke procedural EHRs with regards to nuclear energy prior to the FNA because at that point in time they did not have any real interest in nuclear energy and further had not been invited by TEPCO or any other public institutions to participate in such. The residents, however, have exercised two procedural EHRs with regards to the Fukushima matter since the 2011 disaster. The first of these was the right to information regarding nuclear energy and the accident. Various information has been disclosed by public institutions and TEPCO through the internet, brochures, information centers, and so on, and residents indeed have open access to this information. However, three problems in relation to this access have been identified. These include: (1) reliability—people perceive risks differently to officials, which triggers reliability issues regarding environmental safety; (2) complexity—the older generation, in particular, find this information technical and complex; (3) lack of timely sharing—residents did not receive information about the radioactive contamination in the few first days after the nuclear accident. Secondly, residents have used the right to access to the courts to address two issues, those of poor living conditions and issues of compensation. The analysis indicates that residents have never applied to the court regarding human rights violations in connection with the environment, which indicates that there is little or no social awareness about EHRs. Lastly, we found that the victims of the accident do not generally participate in the decision-making process in any meaningful manner with regards to nuclear energy and the FNA due to the following reasons: (1) people do not have any right to participate in the decision-making process, (2) culture of overwork—a large number of residents work very hard and consequently have very little time for these kinds of activities, (3) residents demanded a referendum rather than public participation, and (4) Japan's patriarchal society—women do not have the opportunity to represent their families.

However, no research has studied the linkage between substantive rights and PRs, which is the gap identified by this book. This research has found empirical evidence that PRs and substantive rights are directly linked in the FNA matter in two ways. Firstly, my study and other similar research shows that the right to information has been violated due to the lack of timely sharing of information and scientifically proven information. This has contributed to health problems in Fukushima because people were exposed to the radiation for longer than necessary, in addition to which they do not find the current radiation levels safe to live in. There is, however, no certainty about the extent to which this has affected people's health because other factors, such as the evacuation process, tsunami, earthquake, and so on, have also affected people's mental health. Some of the environmental impacts of the FNA on people's health may only become clear over longer time periods. Secondly, my research and other similar studies indicate that residents have and do not participate in the decision-making process, particularly with regards to risk-based decisions. This has resulted in the fact that there is no socially acceptable risk level for radioactive contamination. While people perceive the environment to be unsafe, officials claim that there is no serious risk to people's health associated with the radiation levels in the THUs. The lack of consensus over how safe is safe enough has triggered mental health problems in Fukushima because people believe that they are being constantly exposed to radiation in their living environments. This issue can be avoided or eliminated through determining tolerable risk levels in a way that allows the residents to participate in the decision-making process related to the radiation levels. Accordingly, Japan needs to improve access to information and public participation in the decision-making process in order to protect the SHRs guaranteed by its constitution. This research provides empirical evidence that the full realization of fundamental human rights, including the rights to life and health, is dependent on the enjoyment of PRs. In other words, substantive rights and PRs are not alternatives to each other in attempts to achieve a safe environment.

THEORETICAL CONTRIBUTION

This research has critically assessed a number of existing theories regarding EHRs and contributes to the theory by providing empirical insight into the associated theoretical discussion and, indeed, into my understanding of the environmental impacts of the nuclear accident on human rights, including the rights to life, health, and property, and further into the role of human rights, including substantive and procedural EHRs, in addressing environmental issues in connection with the FNA and the formulation of NEP that EHRs theory argues. The theoretical contributions are as follows:

Reinterpretation of Human Rights versus Right to the Environment

Firstly, there has been considerable theoretical discussion over whether a distinct right to the environment should exist since the 1972 Stockholm Declaration. Supporters argue that the environment provides basic human needs including water, air, and food; people cannot exist without a stable environment. Accordingly, human beings have a basic right to be able to live in a clean environment, which could represent an effective means by which to encourage governments to protect the environment. The right to the environment has been recognized by fifty-six countries as constituting a basic human right. Opponents agree that the environment is fundamental to human life and health. However, they also argue that there is no need for a new or distinct human right to the environment because existing human rights, such as the rights to life and health, already require a safe environment in order for them to be enjoyed and realized by individuals in any case. This research contributes to this theoretical discussion as constitutionally recognized human rights in Japan, such as the right to health and property, have been violated by the contaminated environment. If a safe environment is essential to the realization of human rights, this obliges all governments to achieve a safe environment because each nation has a positive obligation to protect human rights which require a safe environment, as evidenced in Fukushima. This means that there may be no need for a distinct right to the environment because existing human rights already require a safe environment as a prerequisite, as evidenced by this study.

Use of Substantive Rights in Environmental Matters

EHRs theory assumes that SHRs can be used by individuals to address issues relating to the environment, and further to affect decisions relating to the environment and nuclear energy. These include the right to the environment and the reinterpretation of human rights. Japan does not recognize the right to the environment, but the country guarantees universal human rights such as the rights to life and health, which can be used by citizens to demand environmental safety. However, my empirical data is not consistent with this theoretical assumption. This research found evidence that Fukushima's residents have never invoked human rights to address the radioactive contamination in the region, though some civil society organizations have demanded the government take action to protect the environment in order to promote human rights in the region. Residents do not invoke human rights in the attempt to oblige the Japanese government to achieve a safe environment, even though the government has a distinct obligation to protect human rights which require a safe environment. This means that the reinterpretation of human rights does not represent an effective route for Fukushima residents to encourage their government to achieve a safe environment and address the contaminated environment. Recognition of human rights within the constitution or the fact that human rights require a safe environment is not in itself sufficient to encourage the public to use existing human rights to achieve a safe environment.

The Linkage between Substantive Rights and PRs

The literature regarding EHRs theory in general discusses substantive rights and PRs largely as being two distinct domains. While some groups have proposed that substantive rights represent an effective means by which to achieve environmental safety, others consider PRs to be an effective tool with which to protect nature. The majority of scholars currently assume that only the right to environment or otherwise a reinterpretation of human rights or PRs would be sufficient to ensure environmental safety. However, this book provides original empirical insight into this theoretical discussion through the finding that PRs and substantive rights are interdependent, meaning that each can only be effective in environmental matters when the other has been fully realized because, as this book shows, the violation of PRs has affected the enjoyment of substantive rights in Fukushima. A safe environment, therefore, can only be realized through the simultaneous enjoyment of substantive rights and PRs.

PRs as Human Rights

The literature regarding EHRs does not generally consider the rights to public participation in decision-making processes or access to information and justice to be human rights because they have already been addressed and recognized by the majority of IEL documents, such as the Aarhus Convention and Rio Declaration. In other words, these three rights are regarded as being procedural rather than human rights, as discussed in chapter 2. Very few scholars consider the fact that these three rights were originally rooted in human rights documents such as the UDHR and ICCPRs. This research provides empirical evidence to the theoretical discussion that SHRs require PRs to be realized and enjoyed. In other words, PRs appear to be complementary to SHRs because they cannot be enjoyed in the absence of PRs.

References

ACHPR. (1987). African charter on human and peoples' rights. Retrieved from http://www.achpr.org/instruments/achpr/

Acton, J. M., & Hibbs, M. (2012). *Why Fukushima was preventable*. Washington, D.C.: Carnegie Endowment for International Peace.

AEC. (2018). Japan atomic energy commission (JAEC). Retrieved from http://www.aec.go.jp/jicst/NC/eng/index.htm

Agger, A. (2011). Towards tailor-made participation: How to involve different types of citizens in participatory governance. *Town Planning Review*, *83(1)*, 29–45.

Ahmedova, A. (2016). Potential threats from transboundary nuclear pollution between Bulgaria and Turkey. *Avrasya Uluslararası Araştırmalar Dergisi*, *4(9)*, 183–191.

Akyuz, E. (2017). Advantages and disadvantages of nuclear energy in Turkey: Public perception. *Eurasian Journal of Environmental Research*, *1(1)*, 1–11.

Albert, B. (1992). Indian lands, environmental policy and military geopolitics in the development of the Brazilian Amazon: The case of the Yanomami. *Development and Change*, *23(1)*, 35–70.

Aldrich, D. P. (2013). Rethinking civil society-state relations in Japan after the Fukushima accident. *Polity*, *45(2)*, 249–264.

Alexander, L. (1998). Are procedural rights derivative substantive rights? *Law and Philosophy*, *17(1)*, 19–42.

Algan, B. (2004). Rethinking third generation human rights. *Ankara Law Review*, *1(1)*, 121–155.

Aliyu, A. S., Ramli, A. T., Garba, N. N., Saleh, M. A., Gabdo, H. T., & Liman, M. S. (2014). Fukushima nuclear accident: Preliminary assessment of the risks to non-human biota. *Radiation Protection Dosimetry*, *163(2)*, 238–250.

Alston, P. (1982). A third generation of solidarity rights: Progressive development or obfuscation of International Human Rights Law? *Netherlands International Law Review*, *29(3)*, 307–322.

Alston, P. (1999). Transplanting foreign norms: Human rights and other international legal norms in Japan. *European Journal of International Law*, *10*, 625–632.

Alston, P. (2013). Introduction: Third generation human rights fact-finding. *ASIL Annual Meeting Proceedings, 107*, 61–62. doi: 10.5305/procannmeetasil.107.0061.

Anderson, M. R., & Boyle, A. E. (1996). *Human rights approaches to environmental protection*. Oxford: Oxford University Press.

ANS. (2005). Chernobyl forum reports 20-year findings, offers recommendations. Retrieved from http://www2.ans.org/pubs/magazines/nn/docs/2005-10-3.pdf

Anton, D. K., & Shelton, D. L. (2011). *Environmental protection and human rights*. Cambridge: Cambridge University Press.

Antonopoulos, I. (2018). The day after: Protecting the human rights affected by environmental challenges after the EU accession to the European Convention on Human Rights. *Environmental Law Review, 20(4)*, 213–224.

Armeni, C. (2016). Participation in environmental decision-making: Reflecting on planning and community benefits for major wind farms. *Journal of Environmental Law, 28(3)*, 415–441.

Arnold, L. (1992). *Windscale 1957: Anatomy of a nuclear accident*. Basingstoke: Macmillan Academic and Professional.

Arnstein, S. R. (1969). A ladder of citizen participation. *Journal of the American Institute of Planners, 35(4)*, 216–224.

Atapattu, S. (2002). The right to a healthy life or the right to die polluted?: The emergence of a human right to a healthy environment under international law. *Tulane Environmental Law Journal, 16(1)*, 65–126.

Babu, R. R. (2012). Constitutional right to property in changing times: The Indian experience. *International Constitutional Law, 6(2)*, 213–247.

Bauer, T. K., Braun, S. T., & Kvasnicka, M. (2017). Nuclear power plant closures and local housing values: Evidence from Fukushima and the German housing market. *Journal of Urban Economics, 99*, 94–106.

Beer, L. W. (2009). *Human rights constitutionalism in Japan and Asia: The writings of Lawrence W. Beer*. Folkestone: Global Oriental.

Beetham, D. (1995). What future for economic and social rights? *Political Studies, 43(1)*, 41–60.

Beierle, T. C. (1998). *Public participation in environmental decisions: An evaluation framework using social goals*. Washington, D.C.: Resources for the Future.

Beierle, T. C. (2010). *Democracy in practice: Public participation in environmental decisions*. New York: Routledge.

Bellamy, R. (2011). Political constitutionalism and the human rights act. *International Journal of Constitutional Law, 9(1)*, 86–111.

Bennett, C. J., & Howlett, M. (1992). The lessons of learning: Reconciling theories of policy learning and policy change. *Policy Sciences, 25(3)*, 275–294.

Bevan, G. (2012). Renewable energy and climate change. *Significance, 9(6)*, 8–12.

Bezdek, R., Ferris, D., Kadri, J., Wolcott, R., Drayton, W., Alley, K., & Larkin, F. (1995). *Environmental justice: Issues, policies, and solutions*. Washington, D.C.: Island Press.

Bier, V. (2001). On the state of the art: Risk communication to the public. *Reliability Engineering & System Safety, 71(2)*, 139–150.

Bishop, P., & Davis, G. (2002). Mapping public participation in policy choices. *Australian Journal of Public Administration, 61(1)*, 14–29.

Blanco, E., & Razzaque, J. (2009). Ecosystem services and human well-being in a globalized world: Assessing the role of law. *Human Rights Quarterly, 31(3)*, 692–720.

Boice, J. D., Jr. (2012). Radiation epidemiology: A perspective on Fukushima. *Journal of Radiological Protection, 32(1)*, 33.

Boling, D. (1998). Access to government-held information in Japan: Citizens' right to know bows to the bureaucracy. *Stanford Journal of International Law, 34*, 1–38.

Bosselmann, K. (2001). Human rights and the environment: Redefining fundamental principles? In *Governing for the environment*. London: Palgrave Macmillan.

Box, M., & McCormack, G. (2004). Terror in Japan: The red army (1969–2001) and aum supreme truth (1987–2000). *Critical Asian Studies, 36(1)*, 91–112.

Boyd, D. R. (2011). *The environmental rights revolution: A global study of constitutions, human rights, and the environment*. Toronto: UBC Press.

Boyd, D. R. (2012). The constitutional right to a healthy environment. *Environment: Science and Policy for Sustainable Development, 54(4)*, 3–15.

Boyle, A. (2007). Human rights or environmental rights? A reassessment. *Fordham Environmental Law Review, 18(3)*, 471–511.

Boyle, A. (2012). Human rights and the environment: Where next? *European Journal of International Law, 23(3)*, 613–642.

Brei, A. T. (2013). Rights & nature. *Journal of Agricultural and Environmental Ethics, 26(2)*, 393–408.

Brems, E., & Adekoya, C. O. (2010). Human rights enforcement by people living in poverty: Access to justice in Nigeria. *Journal of African Law, 54(2)*, 258–282.

Brumfiel, G. (2013). Fallout of fear. *Nature, 493(7432)*, 290–293.

Bunn, M., & Heinonen, O. (2011). Preventing the next Fukushima. *Science, 333(6049)*, 1580–1581.

Busse, M. (2004). Transnational corporations and repression of political rights and civil liberties: An empirical analysis. *Kyklos, 57(1)*, 45–65.

Caglayan, S. (2015). A review of the case law of the European Court of Human Rights in the context of environmental justice. *Hacettepe Hukuk Fakültesi Dergisi, 5*, 81.

Calabrese, E. (2011). Improving the scientific foundations for estimating health risks from the Fukushima incident. *Proceedings of the National Academy of Sciences, 108(49)*, 19447–19448.

Callahan, K. (2007). Citizen participation: Models and methods. *International Journal of Public Administration, 30(11)*, 1179–1196.

Callicott, J. B. (1984). Non-anthropocentric value theory and environmental ethics. *American Philosophical Quarterly, 21(4)*, 299–309.

Camur, E. G. (2017). Civil and political rights vs. social and economic rights: A brief overview. *Bitlis Eren Üniversitesi Sosyal Bilimler Enstitüsü Dergisi, 6(1)*, 205–214.

Cardis, E., & Hatch, M. (2011). The Chernobyl accident—An epidemiological perspective. *Clinical Oncology, 23(4)*, 251–260.

Carpentier, N. (2016). Beyond the ladder of participation: An analytical toolkit for the critical analysis of participatory media processes. *Javnost—The Public, 23(1)*, 70–88.
Cha, V. D., & Kang, D. C. (2018). *Nuclear North Korea: A debate on engagement strategies*. New York: Columbia University Press.
Chanlett-Avery, E. (2010). Japan's nuclear future: Policy debate, prospects, and US interests. Retrieved from https://fas.org/sgp/crs/nuke/RL34487.pdf
Chen, W. M., Kim, H., & Yamaguchi, H. (2014). Renewable energy in eastern Asia: Renewable energy policy review and comparative SWOT analysis for promoting renewable energy in Japan, South Korea, and Taiwan. *Energy Policy, 74*, 319–329.
Chino, M., Nakayama, H., Nagai, H., Terada, H., Katata, G., & Yamazawa, H. (2011). Preliminary estimation of release amounts of 131I and 137Cs accidentally discharged from the Fukushima Daiichi nuclear power plant into the atmosphere. *Journal of Nuclear Science and Technology, 48(7)*, 1129–1134.
Choguill, M. B. G. (1996). A ladder of community participation for underdeveloped countries. *Habitat International, 20(3)*, 431–444.
Christodouleas, J. P., Forrest, R. D., Ainsley, C. G., Tochner, Z., Hahn, S. M., & Glatstein, E. (2011). Short-term and long-term health risks of nuclear-power-plant accidents. *New England Journal of Medicine, 364(24)*, 2334–2341.
Cochran, T. B., Norris, R. S., & Suokko, K. L. (1993). Radioactive contamination at Chelyabinsk-65, Russia. *Annual Review of Energy and the Environment, 18(1)*, 507–528.
Coenen, F. H. J. M. (2009). *Public participation and better environmental decisions: The promise and limits of participatory processes for the quality of environmentally related decision-making*. London, Dordrecht: Springer.
Cohen, J. L., & Arato, A. (1994). *Civil society and political theory*. Cambridge, MA: MIT Press.
Collingsworth, T. (2002). The key human rights challenge: Developing enforcement mechanisms. *Harvard Human Rights Journal, 15*, 183.
Conte, A., & Burchill, R. (2009). *Defining civil and political rights: The jurisprudence of the united nations human rights committee* (2nd/Alex Conte and Richard Burchill, ed.). Aldershot: Ashgate.
Cotton, M. (2009). Evaluating the 'ethical matrix' as a radioactive waste management deliberative decision-support tool. *Environmental Values, 18(2)*, 153–176.
Cotton, M. (2015). Structure, agency and post-Fukushima nuclear policy: An alliance-context-actantiality model of political change. *Journal of Risk Research, 18(3)*, 317–332.
Cotton, M. (2017). Fair fracking? Ethics and environmental justice in United Kingdom shale gas policy and planning. *Local Environment, 22(2)*, 185–202.
Cotton, M. D., & Mahroos-Alsaiari, A. A. (2015). Key actor perspectives on stakeholder engagement in Omani environmental impact assessment: An application of Q-methodology. *Journal of Environmental Planning and Management, 58(1)*, 91–112.
Cramer, B. W. (2009). The human right to information, the environment and information about the environment: From the universal declaration to the Aarhus convention. *Communication Law and Policy, 14(1)*, 73–103.

Crowley, K. D. (1997). Nuclear waste disposal: The technical challenges. *Physics Today, 50(6)*, 32–39.

Cullet, P. (1995). Definition of an environmental right in a human rights context. *Netherlands Quarterly of Human Rights, 13(1)*, 25–40.

Cuttler, J. M., & Pollycove, M. (2009). Nuclear energy and health: And the benefits of low-dose radiation hormesis. *Dose-Response, 7(1)*, 52–89.

Darbishire, H. (2010). Proactive transparency: The future of the right to information? Retrieved from https://elibrary.worldbank.org/doi/abs/10.1596/25031

Davies, S. (2007). In Name or nature? Implementing international environmental procedural rights in the post-Aarhus environment: A Finnish example. *Environmental Law Review, 9(3)*, 190–200.

Davis, S., & Hayes-Conroy, J. (2018). Invisible radiation reveals who we are as people: Environmental complexity, gendered risk, and biopolitics after the Fukushima nuclear disaster. *Social & Cultural Geography, 19(6)*, 720–740.

De Sadeleer, N., Roller, G., & Dross, M. (2005). *Access to justice in environmental matters and the role of NGOs: Empirical findings and legal appraisal*. Groeningen: Europa Law Publishing.

de Saille, S., & Matanle, P. C. (2016). *Fukushima—The triple disaster and its triple lessons: What can be learned about regulation, planning, and communication in an unfolding emergency?* International Network for Government Science Advice. Sheffield: The University of Sheffield.

De Santo, E. M. (2016). Assessing public "participation" in environmental decision-making: Lessons learned from the UK Marine Conservation Zone (MCZ) site selection process. *Marine Policy, 64*, 91–101.

Desgagne, R. (1995). Integrating environmental values into the European Convention on Human Rights. *American Journal of International Law, 89(2)*, 263–294.

DeWit, A., Tetsunari, I., & Masaru, K. (2012). Fukushima and the political economy of power policy in Japan. In *Natural disaster and nuclear crisis in Japan*. London: Routledge.

Dimitrov, R. S. (2016). The Paris agreement on climate change: Behind closed doors. *Global Environmental Politics, 16(3)*, 1–11.

Dobson, A. (1998). *Justice and the environment: Conceptions of environmental sustainability and theories of distributive justice*. Oxford: Oxford University Press.

Domaratzki, Z. (1995). The convention on nuclear safety. *Canadian Nuclear Society Bulletin, 16(1)*, 14–17.

Donnelly, J. (2002). *Universal human rights in theory and practice* (2nd ed.). London, Ithaca, NY: Cornell University Press.

Donnelly, J., & Whelan, D. J. (2017). *International human rights*. London, UK: Hachette.

Downs, J. A. (1992). A healthy and ecologically balanced environment: An argument for a third generation right. *Duke Journal of Comparative & International Law, 3*, 351.

Drenovak-Ivanovic, M. (2015). The development of the right to public participation in environmental matters as a new concept of administrative decision making in Serbia. *Transylvanian Review of Administrative Sciences, 11(44)*, 74–90.

Dreyer, J. (2015). A right to wind: Reconciling environmental human rights and private property rights in Europe. *Int'l L. News*, *44*, 16.

Duan, W., Chen, G., Ye, Q., & Chen, Q. (2011). The situation of hazardous chemical accidents in China between 2000 and 2006. *Journal of Hazardous Materials*, *186(2–3)*, 1489–1494.

Duffield, J. S., & Woodall, B. (2011). Japan's new basic energy plan. *Energy Policy*, *39(6)*, 3741–3749.

Eacott, E. (2001). Clean and healthy environment: The barriers and limitations of this emerging human right. *Dalhousie Journal of Legal Studies*, *10*, 74–101.

ECHR. (2018). Guide on Article 8 of the European convention on human rights. Retrieved from https://www.echr.coe.int/Documents/Guide_Art_8_ENG.pdf

ECHR. (2019). Environment and the European convention on human rights. Retrieved from https://www.echr.coe.int/Documents/FS_Environment_ENG.pdf

Einstein, A. J., Henzlova, M. J., & Rajagopalan, S. (2007). Estimating risk of cancer associated with radiation exposure from 64-slice computed tomography coronary angiography. *JAMA*, *298(3)*, 317–323.

Eisler, R. (2013). *The Fukushima 2011 disaster*. Boca Raton: CRC Press.

Elliott, D. (2013). *Fukushima: Impacts and implications*. Basingstoke: Palgrave Macmillan.

Enserink, B., & Koppenjan, J. (2007). Public participation in China: Sustainable urbanization and governance. *Management of Environmental Quality: An International Journal*, *18(4)*, 459–474.

Etemire, U. (2016). Insights on the UNEP Bali guidelines and the development of environmental democratic rights. *Journal of Environmental Law*, *28(3)*, 393–413.

Etzioni, A. (2010). The normativity of human rights is self-evident. *Human Rights Quarterly*, *32(1)*, 187–197.

EU. (2001). Convention on access to information, public participation in decision-making and access to justice in environmental matters. Retrieved from http://ec.europa.eu/environment/aarhus/

EU. (2018). Doses legal limits. Retrieved from http://www.radioactivity.eu.com/site/pages/Doses_Limits.htm

Eurick, J. P. (1999). The constitutional right to a healthy environment: Enforcing environmental protection through state and federal constitutions. *International Legal Perspectives*, *11*, 185.

EUROPA. (2016). Participatory and procedural rights in environmental matters. Retrieved from http://ec.europa.eu/environment/legal/law/3/module_4_3.htm

EUROPA. (2018). Convention on access to information, public participation in decision-making and access to justice in environmental matters. Retrieved from http://www.unece.org/fileadmin/DAM/env/pp/documents/cep43e.pdf

Faiz, P. M. (2016). The protection of civil and political rights by the constitutional court of Indonesia. *Indonesia Law Review*, *6(2)*, 158–179.

Fam, S. D., Xiong, J., Xiong, G., Yong, D. L., & Ng, D. (2014). Post-Fukushima Japan: The continuing nuclear controversy. *Energy Policy*, *68*, 199–205.

Feldhoff, T. (2018). Visual representations of radiation risk and the question of public (mis-) trust in post-Fukushima Japan. *Societies*, *8(2)*, 32.

Feldman, E. A. (2012). Fukushima: Catastrophe, compensation, and justice in Japan. *DePaul Law Review, 62*, 335.

Figueroa, P. M. (2013). Risk communication surrounding the Fukushima nuclear disaster: An anthropological approach. *Asia Europe Journal, 11(1)*, 53–64.

Fitzmaurice, M. (2002). Some reflections on public participation in environmental matters as a human right in international law. *Non-State Actors and International Law, 2(1)*, 1–22.

Fitzmaurice, M. (2011). The European Court of human rights, environmental damage and the applicability of article 8 of the European Convention on Human Rights and Fundamental Freedoms: Dubetska and others v Ukraine, European Court of Human Rights. *Environmental Law Review, 13(2)*, 107–114.

Forland, A. (1995). *Hope over fear: The establishment of the international atomic energy agency*. Oslo: Institutt for forsvarsstudier.

Freitas, M. P. (2017). Access to environmental justice in Brazil. *International Journal for Court Administration, 8(3)*, 1–6.

French, B., Funamoto, S., Sugiyama, H., Sakata, R., Cologne, J., Cullings, H. M., . . . Preston, D. L. (2018). Population density in Hiroshima and Nagasaki before the bombings in 1945: Its measurement and impact on radiation risk estimates in the life span study of atomic bomb survivors. *American Journal of Epidemiology, 187(8)*, 1623–1629.

Frey, L. M., & Wilhite, K. (2005). Our five basic needs application for understanding the function of behaviour. *Intervention in School and Clinic, 40(3)*, 156–160.

Fujioka, M. (2003). Japan's human rights policy at domestic and international levels: Disconnecting human rights from human security? *Japan Forum, 15(2)*, 287–305.

Fukuda, Y., & Fukuda, K. (2012). Fukushima nuclear power plant accident: Issues on radiation monitoring and its relation to public health. *Journal of Epidemiology and Community Health, 66(12)*, 1083–1084.

Fukushima Prefecture. (2017). Safety of drinking water. Retrieved from http://www.pref.fukushima.lg.jp/site/portal-english/en-25-7.html

Funabashi, Y., & Kitazawa, K. (2012). Fukushima in review: A complex disaster, a disastrous response. *Bulletin of the Atomic Scientists, 68(2)*, 9–21.

Gellers, J. C., & Jeffords, C. (2015). Procedural environmental rights and environmental justice: assessing the impact of environmental constitutionalism. Retrieved from https://ssrn.com/abstract=2646739

Gellers, J. C., & Jeffords, C. (2018). Toward environmental democracy? Procedural environmental rights and environmental justice. *Global Environmental Politics, 18(1)*, 99–121.

Gerteis, C. (2009). *Gender struggles: Wage-earning women and male-dominated unions in post-war Japan*. Harvard: Harvard University Press.

Giviashvili, I. (2011). The implementation of procedural environmental rights: The BTC case study on the implementation of procedural environmental rights (Doctoral dissertation, Nottingham Trent University).

Glazebrook, S. (2009). Human rights and the environment. *Victoria University of Wellington Law Review, 40(1)*, 293–350.

Gleick, P. H. (1996). Basic water requirements for human activities: Meeting basic needs. *Water International*, *21(2)*, 83–92.

GOJ. (2018). Fukushima foods: Safe and delicious. Retrieved from https://www.japan.go.jp/tomodachi/2017/autumn2017/fukushima_food.html

González, Y. M. (2007). Human rights in Japan: Progress and challenges. *Portes: Revista Mexicana De Estudios Sobre La Cuenca Del Pacífico*, *1(2)*, 27–50.

Goodale, M., & Merry, S. E. (Eds.). (2007). *The practice of human rights: Tracking law between the global and the local*. Cambridge, MA: Cambridge University Press.

Goto, A., Bromet, E. J., Ota, M., Ohtsuru, A., Yasumura, S., Fujimori, K., & Pregnancy and Birth Survey Group of the Fukushima Health Management Survey. (2017). The Fukushima nuclear accident affected mothers' depression but not maternal confidence. *Asia Pacific Journal of Public Health*, *29(2)*, 139S–150S.

Goto, A., Rudd, R. E., Bromet, E. J., Suzuki, Y., Yoshida, K., Suzuki, Y., ... & Reich, M. R. (2014). Maternal confidence of Fukushima mothers before and after the nuclear power plant disaster in Northeast Japan: Analyses of municipal health records. *Journal of Communication in Healthcare*, *7(2)*, 106–116.

Guillerm, N., & Cesari, G. (2015). Fighting ambient air pollution and its impact on health: From human rights to the right to a clean environment. *The International Journal of Tuberculosis and Lung Disease*, *19(8)*, 887–897.

Gunderson, R. (2018). Global environmental governance should be participatory: Five problems of scale. *International Sociology*, *33(6)*, 715–737.

Günther, O. (1998). *Environmental information systems*. London, Berlin: Springer.

Gupta, A. (2008). Transparency under scrutiny: Information disclosure in global environmental governance. *Global Environmental Politics*, *8(2)*, 1–7.

Gupta, A. (Ed.). (2010). Transparency in global environmental governance: A coming of age? *Global Environmental Politics*, *10(3)*, 1–9.

Hajjar Leib, L. (2017). *Human rights and the environment: Philosophical, theoretical and legal perspectives*. Leiden: BRILL.

Hallo, R. E. (1996). *Access to environmental information in Europe: The implementation and implications of directive 90/313/EEC*. London, Boston: Kluwer Law International.

Hancock, J. (2003). *Environmental human rights: Power, ethics, and law*. Aldershot: Ashgate.

Harada, K. H., Niisoe, T., Imanaka, M., Takahashi, T., Amako, K., Fujii, Y., & Koizumi, A. (2014). Radiation dose rates now and in the future for residents neighbouring restricted areas of the Fukushima Daiichi Nuclear Power Plant. *Proceedings of the National Academy of Sciences*, *111(10)*, E914–E923. doi: 10.1073/pnas.1315684111.

Hartley, N., & Wood, C. (2005). Public participation in environmental impact assessment-implementing the Aarhus convention. *Environmental Impact Assessment Review*, *25(4)*, 319–340.

Hasegawa, A., Tanigawa, K., Ohtsuru, A., Yabe, H., Maeda, M., Shigemura, J., ... & Ishikawa, T. (2015). Health effects of radiation and other health problems in the aftermath of nuclear accidents, with an emphasis on Fukushima. *The Lancet, 386(9992)*, 479–488.

Hasegawa, K. (2014). The Fukushima nuclear accident and Japan's civil society: Context, reactions, and policy impacts. *International Sociology, 29(4)*, 283–301.

Hassan, D., & Khan, A. (2013). Climate-change-related human rights violations. *Environmental Policy and Law, 43(2)*, 80.

Hatanaka, T., Yoshida, S., Ojino, M., & Ishii, M. (2014). The communication of information such as evacuation orders at the time of a nuclear power station accident: Recommendations for responses by the national government and electric power utilities to the "information disaster". *Japan Medical Association Journal: JMAJ, 57(5–6)*, 293.

Hayao, K. (2014). *The Japanese prime minister and public policy*. Pittsburgh: University of Pittsburgh Press.

Hayashi, M., & Hughes, L. (2013). The policy responses to the Fukushima nuclear accident and their effect on Japanese energy security. *Energy Policy, 59*, 86–101.

Hayward, T. (2005). *Constitutional environmental rights*. Oxford: Oxford University Press.

Hazen, S. (1997). Environmental democracy. *Our Planet, 8(6)*, 31.

Hecht, G., & Callon, M. (2009). *The radiance of France: Nuclear power and national identity after World War II*. London: MIT Press.

Hernandez-Truyol, B. E. (1996). International law, human rights, and LatCrit theory: Civil and political rights: An introduction. *The University of Miami Inter-American Law Review, 28(2)*, 223–243.

Higuchi, O., & Hanita, K. (2017). Do the explanations regarding the safety of Fukushima-produced foods ease consumer's concerns about disease?. *Shinrigaku Kenkyu: The Japanese Journal of Psychology, 88(1)*, 43–50.

Hindmarsh, R. (Ed.). (2013). *Nuclear disaster at Fukushima Daiichi: Social, political and environmental issues*. New York: Routledge.

Hiskes, R. P. (2016). The honour of human rights: Environmental rights and the duty of intergenerational promise. *Human Rights Review, 17(4)*, 463–478.

Hiskes, R. P. (2017). With apologies to the future: Environmental human rights and the politics of communal responsibility. *The International Journal of Human Rights, 21(9)*, 1401–1416.

Hobbs, H., & Trotter, A. (2013). How far have we really come? Civil and political rights in Queensland. *Bond Law Review, 25(2)*, 166–208.

Hölgye, Z., Filgas, R., & Kaizer, J. (1993). Monitoring of the release of transuranium elements in airborne effluents from V-1 nuclear power plant at Jaslovské Bohunice in Czechoslovakia. *Journal of Environmental Radioactivity, 20(3)*, 191–199.

Holifield, R. (2001). Defining environmental justice and environmental racism. *Urban Geography, 22(1)*, 78–90.

Homma, T., & Akimoto, K. (2013). Analysis of Japan's energy and environment strategy after the Fukushima nuclear plant accident. *Energy Policy, 62*, 1216–1225.

Hong, S., Bradshaw, C. J., & Brook, B. W. (2013). Evaluating options for the future energy mix of Japan after the Fukushima nuclear crisis. *Energy Policy*, *56*, 418–424.

House, F. (2014). Freedom in the world 2014: The annual survey of political rights and civil liberties. Retrieved from https://freedomhouse.org/sites/default/files/Freedom%20in%20the%20World%202014%20Booklet.pdf

Howard, R. (1983). The full-belly thesis: Should economic rights take priority over civil and political rights-evidence from sub-Saharan Africa. *Human Rights Quarterly*, *5*, 467.

HRN. (2019). Written statement submitted by human rights now, a non-governmental organization in special consultative status. Retrieved from http://hrn.or.jp/eng/wp-content/uploads/2019/02/3668_A_HRC_40_NGO_Sub_En.pdf

Hultquist, C., & Cervone, G. (2018). Citizen monitoring during hazards: Validation of Fukushima radiation measurements. *Geo Journal*, *83(2)*, 189–206.

Humby, T. (2010). The biowatch case: Major advance in South African law of costs and access to environmental justice. *Journal of Environmental Law*, *22(1)*, 125–134.

Humphreys, S., & Robinson, M. (Eds.). (2010). *Human rights and climate change*. Cambridge, MA: Cambridge University Press.

IAE. (2016). *Energy policies of IEA countries: Japan*. Paris: OECD.

IAEA. (2016). The international nuclear and radiological event scale. Retrieved from http://www-ns.iaea.org/tech-areas/emergency/ines.asp

IAEA. (2018). The atomic energy basic law. Retrieved from https://inis.iaea.org/search/search.aspx?orig_q=RN:12597062

Iguchi, M., & Koga, M. (2015). The state of environment and energy governance in Japan. *Environmental Challenges and Governance: Diverse Perspectives from Asia*, *37*, 219.

Iida, K. (2004). Human rights and sexual abuse: The impact of international human rights law on Japan. *Human Rights Quarterly*, *26(2)*, 428–453.

Irie, K. (2011). Problems and legislative remedies of the parallel law systems in Japan for nuclear power reactors. *Transactions of the Atomic Energy Society of Japan*, *10(1)*, 30–47.

IRSN. (2016). Health consequences of the Fukushima accident. Retrieved from https://www.irsn.fr/EN/Pages/home.aspx

Ishay, M. (2008). *The history of human rights: From ancient times to the globalization era* (New ed.). London, Berkeley, CA: University of California Press.

Ishida, M., & Yamazaki, H. (2017). Radioactive contamination in the Tokyo metropolitan area in the early stage of the Fukushima Daiichi Nuclear Power Plant (FDNPP) accident and its fluctuation over five years. *PLoS One*, *12(11)*, 1–18.

Ishikawa, T. (2017). Radiation doses and associated risk from the Fukushima Nuclear Accident: A review of recent publications. *Asia Pacific Journal of Public Health*, *29(2)*, 18S–28S.

Ishiyama, F. I., & Kitayama, A. (1994). Overwork and career-centered self-validation among the Japanese: Psychosocial issues and counselling implications. *International Journal for the Advancement of Counselling*, *17(3)*, 167–182.

Ito, S., Goto, A., Ishii, K., Ota, M., Yasumura, S., Fujimori, K., & Pregnancy and Birth Survey Group of the Fukushima Health Management Survey. (2017). Fukushima mothers' concerns and associated factors after the Fukushima nuclear power plant disaster: Analysis of qualitative data from the Fukushima Health Management Survey, 2011 to 2013. *Asia Pacific Journal of Public Health*, *29(2l)*, 151S–160S.

Iwasawa, Y. (1998). *International law, human rights, and Japanese law: The impact of international law on Japanese law*. Oxford: Clarendon.

JAEC. (2015). Utilization of nuclear power in Japan. Retrieved from http://www.aec.go.jp/jicst/NC/about/kettei/151201.pdf

JAEC. (2017). The basic policy for nuclear energy 2017. Retrieved from http://www.aec.go.jp/jicst/NC/eng/basic_policy.htm

Jankowitsch, O., & Tonhauser, W. (1997). The convention on nuclear safety. *Austrian Review of International and European Law Online*, *2(1)*, 319–340.

Jargin, S. V. (2011). Thyroid cancer after Chernobyl: Obfuscated truth. *Dose-Response*, *9(4)*, 471.

Jesenko, K. (2018). Freedom of expression in environmental cases before the European Court of Human Rights. *ERA Forum*, *19(2)*, 295–305.

Jimba, M., Kimura, S., Tsuboi, M., Komasa, Y., & Hasebe, A. K. (2013). Prioritizing mental health issues of community residents affected by Fukushima Daiichi nuclear power plant accident. *Clinical Neuropsychiatry: Journal of Treatment Evaluation*, *10(6)*, 241.

Johnson, P. L. (2013). The post-Fukushima Daiichi response: The role of the convention on nuclear safety in strengthening the legal framework for nuclear safety. *Nuclear Law Bulletin*, *2013(1)*, 7–21. Paris: OECD Publishing.

Johnston, B. R. (1995). Human rights and the environment. *Human Ecology*, *23(2)*, 111–123.

Johnston, B. R. (2013). Human needs and environmental rights to water: A biocultural systems approach to hydro development and management. *Ecosphere*, *4(3)*, 39–15.

Joskow, P. L., & Parsons, J. E. (2012). The future of nuclear power after Fukushima. *Economics of Energy & Environmental Policy*, *1(2)*, 99–114.

Jung, C., Hirschl, R., & Rosevear, E. (2014). Economic and social rights in national constitutions. *The American Journal of Comparative Law*, *62(4)*, 1043–1094.

Kaletski, E., Minkler, L., Prakash, N., & Randolph, S. (2016). Does constitutionalizing economic and social rights promote their fulfillment? *Journal of Human Rights*, *15(4)*, 433–453.

Kamiya, K., Ozasa, K., Akiba, S., Niwa, O., Kodama, K., Takamura, N., ... & Wakeford, R. (2015). Long-term effects of radiation exposure on health. *The Lancet*, *386(9992)*, 469–478.

Kamminga, M. T. (1995). The IAEA convention on nuclear safety. *The International and Comparative Law Quarterly*, *44(4)*, 872–882.

KANTEI. (2018). The constitution of Japan. Retrieved from https://japan.kantei.go.jp/constitution_and_government_of_japan/constitution_e.html

Kasemir, B. (2003). *Public participation in sustainability science: A handbook.* Cambridge, MA: Cambridge University Press.

Kasperson, J. X., & Kasperson, R. E. (2001). *Global environmental risk.* London, Tokyo: United Nations University Press.

Kavanagh, A. (2009). *Constitutional review under the UK Human Rights Act.* Cambridge, MA: Cambridge University Press.

Kawaguchi, D., & Yukutake, N. (2017). Estimating the residential land damage of the Fukushima nuclear accident. *Journal of Urban Economics, 99*, 148–160.

Kawano, Y., Shepard, D., Shobugawa, Y., Goto, J., Suzuki, T., Amaya, Y., ... & Takahashi, T. (2012, October). A map for the future: measuring radiation levels in Fukushima, Japan. Paper presented at the GHTC, 53–58. doi: 10.1109/GHTC.2012.18.

Keles, R., Hamamcı, C., & Çoban, A. (2012). *Çevre politikası.* Ankara: İmge Kitabevi.

Kermisch, C., & Labeau, P. (2013). Communicating about nuclear events: Some suggestions to improve INES. *Reliability Engineering and System Safety, 119*, 165–171.

Kermisch, C. F. N., & Taebi, B. (2017). Sustainability, ethics and nuclear energy: Escaping the dichotomy. *Sustainability, 9(3)*, 446.

Khan, J. (2018). K. Post-disaster policy decision making and the prospects of human rights—The case of Fukushima Daiichi nuclear. *Sociology and Anthropology, 6(1)*, 116–134.

Kingston, J. (2012). *Natural disaster and nuclear crisis in Japan: Response and recovery after Japan's 3/11.* London: Routledge.

Kingston, J. (2013a). *Critical issues in contemporary Japan.* London: Routledge.

Kingston, J. (2013b). Nuclear power politics in Japan, 2011–2013. *Asian Perspective, 37(4)*, 501–521.

Kingston, J. (2014a). *Critical issues in contemporary Japan.* London: Routledge.

Kingston, J. (2014b). Mismanaging risk and the Fukushima nuclear crisis. *The Asia-Pacific Journal, 10(12)*, 39.

Kiss, A. C., & Dominick, M. F. (1980). The international legal significance of the human rights provisions of the Helsinki final act. *Vanderbilt Journal of Transnational Law, 13(2–3)*, 293.

Kiss, G. (2014). Why should the public participate in environmental decision-making? Theoretical arguments for public participation. *Periodica Polytechnica Social and Management Sciences, 22(1)*, 13–20.

Knox, J. H. (2009). Linking human rights and climate change at the United Nations. *Harvard Environmental Law Review, 33*, 477.

Komiya, N. (1999). A cultural study of the low crime rate in Japan. *British Journal of Criminology, 39(3)*, 369–390.

Komiyama, R., & Fujii, Y. (2017). Assessment of post-Fukushima renewable energy policy in Japan's nation-wide power grid. *Energy Policy, 101*, 594–611.

Konisky, M., Thomas, C., & Beierle, D. (2001). Innovations in public participation and environmental decision making: Examples from the Great Lakes region. *Society &Natural Resources, 14(9)*, 815–826.

Kovar, J. D. (1993). A short guide to the Rio declaration. *Colorado Journal of International Environmental Law and Policy, 4(1)*, 118–140.

Koyama, K. (2013). Japan's Post-Fukushima energy policy challenges. *Asian Economic Policy Review, 8(2)*, 274–293.

Kravchenko, S. (2010a). Is access to environmental information a fundamental human right? *Oregon Review of International Law, 11(2)*, 1–40.

Kravchenko, S. (2010b). Procedural rights as a crucial tool to combat climate change. *Georgia Journal of International and Comparative Law, 38(3)*, 613.

Krstić, I., & Čučković, B. (2015). Procedural aspects of article 8 of the ECHR in environmental cases: The greening of human rights law. *Anali Pravnog fakulteta u Beogradu, 63(3)*, 170–189.

Kukihara, H., Yamawaki, N., Uchiyama, K., Arai, S., & Horikawa, E. (2014). Trauma, depression, and resilience of earthquake/tsunami/nuclear disaster survivors of Hirono, Fukushima, Japan. *Psychiatry and Clinical Neurosciences, 68(7)*, 524–533.

Kunii, Y., Suzuki, Y., Shiga, T., Yabe, H., Yasumura, S., Maeda, M., ... & Abe, M. (2016). Severe psychological distress of evacuees in evacuation zone caused by the Fukushima Daiichi nuclear power plant accident: The Fukushima health management survey. *PLoS One, 11(7)*, 1–15.

Kurosaki, A. (2017). Nuclear energy and nuclear-weapon potential: A historical analysis of Japan in the 1960s. *The Non-Proliferation Review, 24(1–2)*, 47–65.

Lee, M., & Abbot, C. (2003). The usual suspects? Public participation under the Aarhus Convention. *The Modern Law Review, 66(1)*, 80–108.

Lemos, M. C., & Agrawal, A. (2006). Environmental governance. *Annual Review of Environment and Resources, 31*, 297–325.

Lenzen, M. (2008). Life cycle energy and greenhouse gas emissions of nuclear energy: A review. *Energy Conversion And Management, 49(8)*, 2178–2199.

Lewis, B. (2012a). Environmental rights or a right to the environment?: Exploring the nexus between human rights and environmental protection. *Macquarie Journal of International and Comparative Environmental Law, 8(1)*, 36–47.

Lewis, B. (2012b). Human rights and environmental wrongs: Achieving environmental justice through human rights law. *International Journal for Crime, Justice and Social Democracy, 1(1)*, 65–73.

Lim, J., Petersen, S., Schwarz, D., Schwarz, R., & Maru, D. (2013). A rights-based approach to indoor air pollution. *Health and Human Rights, 15(2)*, 160–167.

Lorenzo, J. A. P. (2015). Investment promotion as progressive realization of economic and social rights. *Asia-Pacific Journal on Human Rights and the Law, 16(1–2)*, 55–103.

Lührs, N., Jager, N. W., Challies, E., & Newig, J. (2018). How participatory should environmental governance be? Testing the applicability of the Vroom-Yetton-Jago model in public environmental decision-making. *Environmental Management, 61(2)*, 249–262.

Lundgren, R. E., & McMakin, A. H. (2018). *Risk communication: A handbook for communicating environmental, safety, and health risks*. Canada: John Wiley & Sons.

Lynn, F. M. (1990). Public participation in risk management decisions: The right to define, the right to know, and the right to act. *Risk: Issues in Health & Safety, 1(2)*, 95.

Lytton, C. H. (2000). Environmental human rights: Emerging trends in international law and ecocide. *Environmental Claims Journal, 13(1)*, 73–91.

Maeda, M., & Oe, M. (2017). Mental health consequences and social issues after the Fukushima disaster. *Asia Pacific Journal of Public Health, 29(2)*, 36–46.

Mah, D. N. Y., Hills, P., & Tao, J. (2014). Risk perception, trust and public engagement in nuclear decision-making in Hong Kong. *Energy Policy, 73*, 368–390.

Mahaffey, J. A. (2015). *Atomic accidents: A history of nuclear meltdowns and disasters: From the Ozark mountains to Fukushima.* New York, NY: Pegasus Books.

Maier, K. (2001). Citizen participation in planning: Climbing a ladder?. *European Planning Studies, 9(6)*, 707–719.

Maki, J. M. (1990). The constitution of Japan: Pacifism, popular sovereignty, and fundamental human rights. *Law and Contemporary Problems, 53(1)*, 73–87.

Marris, C., Langford, I. H., & Riordan, T. O. (1996). Integrating sociological and psychological approaches to public perceptions of environmental risks: Detailed results from a questionnaire survey (CSERGE Working Paper GEC 96-07). University of East Anglia, ISSN 0967-8875.

Maslow, A. H. (1943). A theory of human motivation. *Psychological Review, 50(4)*, 370.

Mason, M. (2010). Information disclosure and environmental rights: The Aarhus Convention. *Global Environmental Politics, 10(3)*, 10–31.

Mason, M. (2012). *Environmental democracy: A contextual approach.* London: Routledge.

May, J. (2013). Constitutional directions in procedural environmental rights. *Journal of Environmental Law & Litigation, 28*, 27.

May, M. M. (2017). Safety first: The future of nuclear energy outside the United States. *Bulletin of the Atomic Scientists, 73(1)*, 38–43.

McCombie, C., & Jefferson, M. (2016). Renewable and nuclear electricity: Comparison of environmental impacts. *Energy Policy, 96*, 758–769.

McCormack, G., Box, M., & Truth, S. (2004). Terror in Japan. *The Asia-Pacific Journal, 2(6)*, 1–17.

McGoldrick, D. (1996). Sustainable development and human rights: An integrated conception. *International & Comparative Law Quarterly, 45(4)*, 796–818.

McGowan, F. (1991). Controlling the greenhouse effect on the role of renewables. *Energy Policy, 19(2)*, 110–118.

McKinney, M., & Harmon, W. (2002). Public participation in environmental decision making: Is it working?. *National Civic Review, 91(2)*, 149–170.

McLellan, B. C. (2017). Nuclear power and energy policy in Japan: A foreign resident energy researcher's perspective on recent progress. *International Electronic Journal of Nuclear Safety and Simulation, 8(2)*, 85–90.

McNeill, D. (2012). The Fukushima nuclear crisis and the fight for compensation. *The Asia-Pacific Journal, 10(10)*, 1–11.

Meltzer, J. (2011). *After Fukushima: What's next for Japan's energy and climate change policy?* Washington, D.C.: Brookings Institution.
METI. (2015). Long-term energy supply and demand outlook. Retrieved from http://www.meti.go.jp/english/press/2015/pdf/0716_01a.pdf
METI. (2017). Japan's energy white paper 2017. Retrieved from http://www.meti.go.jp/english/report/downloadfiles/energy_hakusho_201711.pdf
Meyer, L. H. (2017). *Intergenerational justice*. London: Routledge.
Michener, W. K., Brunt, J. W., & Stafford, S. G. ([1952] 1994). *Environmental information management and analysis: Ecosystem to global scales*. London: Taylor & Francis.
Millner, F. (2011). Access to environmental justice. *Deakin Law Review, 16(1)*, 189–207.
Mindell, D. P. (2009). Humans need biodiversity. *Science, 323(5921)*, 1562–1563.
Mizuno, T., & Kubo, H. (2013). Overview of active cesium contamination of freshwater fish in Fukushima and Eastern Japan. *Scientific Reports, 3(1)*, 1742–1746.
MOE. (2019). Japan's regulations and environmental law. Retrieved from http://www.env.go.jp/en/coop/pollution.html
MOFA. (2008). National report by Japan for the universal periodic review. Retrieved from https://www.mofa.go.jp/policy/human/report0803.pdf
MOFA. (2009). Japan's human rights commitments and pledges. Retrieved from https://www.mofa.go.jp/files/000175306.pdf
MOFA. (2013). Japan's human rights commitments and pledges. Retrieved from https://www.mofa.go.jp/policy/human/pledge1109.html
MOFA. (2017). Candidate for the Human Rights Council 2017–2019. Retrieved from https://www.mofa.go.jp/files/000175485.pdf
Montjoie, M. (2016). Treaty implementation applied to conventions on nuclear safety. *Nuclear Law Bulletin, 2015(2)*, 9–34. Paris: OECD Publishing.
Monzen, S., Hosoda, M., Tokonami, S., Osanai, M., Yoshino, H., Hosokawa, Y., ... & Kashiwakura, I. (2011). Individual radiation exposure dose due to support activities at safe shelters in Fukushima Prefecture. *PLoS One, 6(11)*, e27761.
Mousseau, T. A., & Møller, A. P. (2014). Genetic and ecological studies of animals in Chernobyl and Fukushima. *Journal of Heredity, 105(5)*, 704–709.
Munro, A. (2018). Economic valuation of a rise in ambient radiation risks: Hedonic pricing evidence from the accident in Fukushima. *Journal of Regional Science, 58(3)*, 635–658.
Murakami, M., Nakatani, J., & Oki, T. (2016). Evaluation of risk perception and risk-comparison information regarding dietary radionuclides after the 2011 Fukushima nuclear power plant accident. *PLoS One, 11(11)*, 1–22.
Murakami, M., & Oki, T. (2012). Estimation of thyroid doses and health risks resulting from the intake of radioactive iodine in foods and drinking water by the citizens of Tokyo after the Fukushima nuclear accident. *Chemosphere, 87(11)*, 1355–1360.
Murakami, M., Ono, K., Tsubokura, M., Nomura, S., Oikawa, T., Oka, T., … & Oki, T. (2015). Was the risk from nursing-home evacuation after the Fukushima accident higher than the radiation risk? *PLoS One, 10(9)*, e0137906. doi: 10.1371/journal.pone.0137906.

Murakami, M., Tsubokura, M., Ono, K., Nomura, S., & Oikawa, T. (2017). Additional risk of diabetes exceeds the increased risk of cancer caused by radiation exposure after the Fukushima disaster. *PLoS One, 12(9)*, e0185259.

Murray, R. L.(2009). *Nuclear energy: An introduction to the concepts, systems, and applications of nuclear processes* (6th ed.). Amsterdam, Boston: Butterworth-Heinemann/Elsevier.

Nakamura, H. (2017). Political and environmental attitude toward participatory energy and environmental governance: A survey in post-Fukushima Japan. *Journal of Environmental Management, 201*, 190–198.

Nakamura, H. (2018). Willingness to know and talk: Citizen attitude toward energy and environmental policy deliberation in post-Fukushima Japan. *Energy Policy, 115*, 12–22.

Nakanishi, Y. (2017). *Contemporary issues in human rights law: Europe and Asia*. Singapore: Springer.

Neary, I. (2002). *Human rights in Japan, South Korea and Taiwan*. London: Routledge.

Nickel, J. W. (1993). The human right to a safe environment: Philosophical perspectives on its scope and justification. *The Yale Journal of International Law, 18(1)*, 281.

Nikiforov, Y. E. (2006). Radiation-induced thyroid cancer: What we have learned from Chernobyl. *Endocrine Pathology, 17(4)*, 307–318.

Nishino, T., Ouchi, M., Tsuburaya, S. I., Tanaka, T., & Hokugo, A. (2011). Emergency evacuation of Fukushima residents living in the vicinity of nuclear power station. In *Proceedings of the international symposium on engineering lessons learned from the 2011 great East Japan earthquake, 1692–1703*. March 1–4, 2012, Tokyo, Japan.

North, S. (2011). Deadly virtues: Inner-worldly asceticism and karôshi in Japan. *Current Sociology, 59(2)*, 146–159.

NPA. (2018). Act on prevention of radiation hazards due to radioisotopes, etc. and its relevant ordinances. Retrieved from http://www.nsr.go.jp/english/library/RI.html

NRA. (2013). Enforcement of the new regulatory requirements for commercial nuclear power reactors. Retrieved from http://www.nsr.go.jp/data/000067212.pdf

NRA. (2015). Annual report. Retrieved from http://www.nsr.go.jp/data/000242168.pdf

NRA. (2016). Convention on nuclear safety national report of Japan for the seventh review meeting. Retrieved from https://www-ns.iaea.org/downloads/ni/safety_convention/7th-review-meeting/japan_nr-7th-rm.pdf

NRC. (2018). Subpart D—Radiation dose limits for individual members of the public. Retrieved from https://www.nrc.gov/reading-rm/doc-collections/cfr/part020/part020-1301.html

NSR. (2018). Radiation hazards prevention act. Retrieved from http://www.nsr.go.jp/data/000182245.pdf

Oberthür, S., Buck, M., Müller, S., Pfahl, S., Tarasofsky, R. G., Werksman, J., & Palmer, A. (2002). *Participation of non-governmental organisations in international environmental governance: Legal basis and practical experience.* Berlin: Ecologic.

OECD. (2006). Law on the peaceful uses of nuclear energy: Key concepts. *Nuclear Law Bulletin, 2005(2).* Paris: OECD Publishing. doi: 10.1787/nuclear_law-2005-5k9czgt8rs0q.

OECD. (2017). Nuclear legislation in OECD and NEA countries Japan. Retrieved from https://www.oecd-nea.org/law/legislation/japan.html

OECD, & NEA. (2006). *Nuclear legislation in OECD countries: Japan.* Paris: OECD Publishing.

OHCHR. (1966). International covenant on civil and political rights. Retrieved from https://www.ohchr.org/EN/ProfessionalInterest/Pages/CCPR.aspx

OHCHR. (1976). International covenant on civil and political rights. Retrieved from https://www.ohchr.org/EN/ProfessionalInterest/Pages/CCPR.aspx

OHCHR. (2013). Report of the Special Rapporteur on the right of everyone to the enjoyment of the highest attainable standard of physical and mental health, Anand Grover. Retrieved from https://www.ohchr.org/Documents/HRBodies/HRCouncil/RegularSession/Session23/A-HRC-23-41-Add3_en.pdf

OHCHR. (2014). Concluding observations on the sixth periodic report of Japan. Retrieved from http://docstore.ohchr.org/SelfServices/FilesHandler.ashx?enc=6QkG1d%2FPPRiCAqhKb7yhsuBJT%2Fi29ui%2Fb4Ih9%2FUIJO9nQa93Boy0croOoLTDvEPGY0kpztyF26TNPPD6smh3p9YJ5KgXGu0vYZb1NM8mpET5PRv%2FLCx0HP6sZ3QjgcWI

Oliver, P. (2013). Access to information and to justice in EU environmental law: The Aarhus convention. *Fordham International Law Journal, 36(5)*, 1423.

Omoto, A. (2013). The accident at TEPCO's Fukushima-Daiichi Nuclear Power Station: What went wrong and what lessons are universal? *Nuclear Instruments and Methods in Physics Research Section A: Accelerators, Spectrometers, Detectors and Associated Equipment, 731*, 3–7.

Orji, U. J. (2012). Nigeria: Right to a clean environment—Some reflections. *Environmental Policy and Law, 42(4)*, 285–293.

Osaka, E. (2012). Corporate liability, government liability, and the Fukushima nuclear disaster. *Pacific Rim Law & Policy Journal, 21*, 433.

Otsuki, K. (2016). Procedural equity and corporeality: Imagining a just recovery in Fukushima. *Journal of Rural Studies, 47*, 300–310.

Padilla, E. (2002). Intergenerational equity and sustainability. *Ecological Economics, 41(1)*, 69–83.

Palerm, J. R. (1999). Public participation in environmental decision making: Examining the Aarhus convention. *Journal of Environmental Assessment Policy and Management, 1(2)*, 229–244.

Pallemaerts, M. (1992). International environmental law from Stockholm to Rio: Back to the future? *Review of European Community & International Environmental Law, 1(3)*, 254–266.

Parker, M. R. (2009). Two sides of the same coin: Considering human rights when designing environmental policy. *Journal of Land Use & Environmental Law, 25(1)*, 109–136.

Pascan, A. (2012). Considerations on the limitation of the right to property. *Journal of Danubian Studies and Research, 2(1)*, 78–85.

Pathak, P. (2014). Human Rights approach to environmental protection. *OIDA International Journal of Sustainable Development, 7(1)*, 17–24.

Peeters, M., & Nóbrega, S. (2014). Climate change-related Aarhus conflicts: How successful are procedural rights in EU climate law? *Review of European, Comparative & International Environmental Law, 23(3)*, 354–366.

Pelzer, N. (2009). Nuclear new build. *Nuclear Law Bulletin*, *2*, 5–21.

Perrow, C. (1994). The limits of safety: The enhancement of a theory of accidents. *Journal of Contingencies and Crisis Management, 2(4)*, 212–220.

Peters, B. (2017). Unpacking the diversity of procedural environmental rights: The European convention on human rights and the Aarhus convention. *Journal of Environmental Law, 30(1)*, 1–27.

Pierce, R. (2008). Evaluating information: Validity, reliability, accuracy, triangulation. In *Research methods in politics: A practical guide*. Canada: Sage Publications.

Pierpoint, L. (2011). Fukushima, Facebook and feeds: Informing the public in a digital era. *The Electricity Journal, 24(6)*, 53–58.

Piovesan, F. (2004). Social, economic and cultural rights and civil and political rights. *Sur Revista Internacional de Direitos Humanos, 1(1)*, 20–47.

Pirro, D. E. (2008). Access to environmental justice. *Environmental Policy and Law*, *38*, 272.

Port, K. L. (1991). The Japanese international law “revolution”: International human rights law and its impact in Japan. *Stanford Journal of International Law, 28(1)*, 139.

Pretty, J. N. (2007). *The SAGE handbook of environment and society*. Los Angeles: SAGE.

Proso, M. (2016). Civil protection of the human right to a safe and healthy environment. economic and social development: Book of proceedings, 282–288.

Quaschning, V. (2009). *Renewable energy and climate change*. GB: Wiley.

Rajan, S. R. (2011). Classical environmentalism and environmental human rights: An exploration of their ontological origins and differences. *Journal of Human Rights and the Environment*, *2*, 106.

Ramkumar, V., & Petkova, E. (2007). Transparency and environmental governance. The right to know: Transparency for an open world. Retrieved from http://policydialogue.org/files/publications/Transparency_in_Env_Governance_Ramkumar_Petkova.pdf

Reader, S. (2006). Does a basic needs approach need capabilities? *Journal of Political Philosophy, 14(3)*, 337–350.

Rehdanz, K., Welsch, H., Narita, D., & Okubo, T. (2015). Well-being effects of a major natural disaster: The case of Fukushima. *Journal of Economic Behavior & Organization, 116*, 500–517.

Renn, O. (1998). The role of risk perception for risk management. *Reliability Engineering & System Safety, 59(1)*, 49–62.

Renn, O. (2015). Stakeholder and public involvement in risk governance. *International Journal of Disaster Risk Science, 6(1)*, 8–20.

Renteln, A. D. (2013). *International human rights: Universalism versus relativism*. New Orleans: Quid Pro Books.

Repeta, L. (1987). The international covenant on civil and political rights and human rights law in Japan. *Law Japan, 20(1)*, 1.

Rhode, D. L. (2004). *Access to justice*. Oxford: Oxford University Press.

Rieu, A. M. (2013). Thinking after Fukushima: Epistemic shift in social sciences. *Asia Europe Journal, 11(1)*, 65–78.

Risse-Kappen, T., Risse, T., Ropp, S. C., & Sikkink, K. (Eds.). (1999). *The power of human rights: International norms and domestic change*. Cambridge, MA: Cambridge University Press.

Rodriguez Rivera, L. E. (2001). Is the human right to environment recognized under international law? It depends on the source. *Colorado Journal of International Environmental Law and Policy, 12(1)*, 1.

Roeser, S. (2011). Nuclear energy, risk, and emotions. *Philosophy & Technology, 24(2)*, 197–201.

Rogers, G. O. (1998). Siting potentially hazardous facilities: What factors impact perceived and acceptable risk? *Landscape and Urban Planning, 39(4)*, 265–281.

Rose-Ackerman, S., & Halpaap, A. A. (2002). The Aarhus Convention and the politics of process: The political economy of procedural environmental rights. In *An introduction to the law and economics of environmental policy: Issues in institutional design*. Bingley: Emerald Group Publishing Limited.

Rothstein, L. (2001). Tokai "downwinders" speak out (nuclear accident in Tokaimura, Japan). *Bulletin of the Atomic Scientists, 57(5)*, 11.

Rowe, G., & Frewer, L. J. (2000). Public participation methods: A framework for evaluation. *Science, Technology, & Human Values, 25(1)*, 3–29.

Ruppel, O. C. (2008). Third-generation human rights and the protection of the environment in Namibia. Human rights and the rule of law in Namibia. Retrieved from http://www.the-eis.com/data/literature/Third_generation%20human%20rights%20and%20the%20protection.pdf

Sacchi, S., Riva, P., & Brambilla, M. (2013). When mother earth rises up: Anthropomorphizing nature reduces support for natural disaster victims. *Social Psychology, 44(4)*, 271–277.

Sachs, W. (2004). Environment and human rights. *Development, 47(1)*, 42–49.

Saenko, V., Ivanov, V., Tsyb, A., Bogdanova, T., Tronko, M., Demidchik, Y., & Yamashita, S. (2011). The Chernobyl accident and its consequences. *Clinical Oncology, 23(4)*, 234–243.

Samy, M., Lokpobiri, H., & Dawodu, A. (2015). The arguments for environmental rights in Niger Delta: A theoretical framework. *Society and Business Review, 10(2)*, 132–149.

Sato, A., & Lyamzina, Y. (2018). Diversity of concerns in recovery after a nuclear accident: A perspective from Fukushima. *International Journal of Environmental Research and Public Health, 15(2)*, 350.

Schlosberg, D. (2009). *Defining environmental justice: Theories, movements, and nature*. Oxford: Oxford University Press.

Schmid, E., & Nolan, A. (2014). 'Do no harm'? Exploring the scope of economic and social rights in transitional justice. *International Journal of Transitional Justice, 8(3)*, 362–382.

Schreurs, M. A., & Miranda, A. (2014). The ethics of nuclear energy: Germany's energy politics after Fukushima. *The Journal of Social Science, 77*, 9–29.

Schwelb, E. (1968). Civil and political rights: The international measures of implementation. *American Journal of International Law, 62(4)*, 827–868.

Scott, I. K. (2000). The inter-American system of human rights: An effective means of environmental protection? *Virginia Environmental Law Journal*, 197–237.

Semenov, B. A. (1983). Nuclear power in the Soviet Union. *IAEA Bulletin, 25(2)*, 47–59.

Sen, A. (2017). Elements of a theory of human rights. In *Justice and the capabilities approach*. London: Routledge.

Sexton, K. (2013). Evolution of public participation in the assessment and management of environmental health risks: A brief history of developments in the United States. *Journal of Public Health Research, 2(2)*, 106–112.

Shadrina, E. (2012). Fukushima fallout: Gauging the change in Japanese nuclear energy policy. *International Journal of Disaster Risk Science, 3(2)*, 69–83.

Shelton, D. (1991). Human rights, environmental rights, and the right to environment. *Stanford Journal of International Law, 28*, 103.

Shelton, D. (2002). Human rights and the environment: Jurisprudence of human rights bodies. *Environmental Policy and Law, 32*, 158.

Shelton, D. (2006). Human rights and the environment: What specific environmental rights have been recognized? *Denver Journal of International Law & Policy, 35*, 129.

Shelton, D. (2008a). Environmental rights and Brazil's obligations in the inter-American human rights system. *The George Washington International Law Review, 40*, 733.

Shelton, D. (2008b). Human rights and the environment: Problems and possibilities. *Environmental Policy and Law, 38(1/2)*, 41–49.

Shimada, Y., Nomura, S., Ozaki, A., Higuchi, A., Hori, A., Sonoda, Y., … Tsubokura, M. (2018). Balancing the risk of the evacuation and sheltering-in-place options: A survival study following Japan's 2011 Fukushima nuclear incident. *BMJ Open, 8(7)*, 1–9.

Shinohara, N., Tokumura, M., Kazama, M., Yoshino, H., Ochiai, S., & Mizukoshi, A. (2013). Indoor air quality, air exchange rates, and radioactivity in new built temporary houses following the great east Japan earthquake in Minamisoma, Fukushima. *Indoor Air, 23(4)*, 332–341.

Shozugawa, K., Riebe, B., Walther, C., Brandl, A., & Steinhauser, G. (2016). Fukushima-derived radionuclides in sediments of the Japanese Pacific Ocean coast and various Japanese water samples (seawater, tap water, and coolant water of Fukushima Daiichi Reactor Unit 5). *Journal of Radioanalytical and Nuclear Chemistry, 307(3)*, 1787–1793.

Shrader-Frechette, K. (2012). Nuclear catastrophe, disaster-related environmental injustice, and Fukushima, Japan: Prima-facie evidence for a Japanese "Katrina". *Environmental Justice, 5(3)*, 133–139.

Shrivastava, K. M. (2009). *The right to information*. Atlanta: Lancer Publishers LLC.

Simons, J. A., Irwin, D. B., & Drinnien, B. A. (1987). Maslow's hierarchy of needs. *October, 9(2009)*, 222.

Sjöberg, L. (2000). Factors in risk perception. *Risk Analysis, 20(1)*, 1–12.

Sjöberg, L., Moen, B. E., & Rundmo, T. (2004). Explaining risk perception. An evaluation of the psychometric paradigm in risk perception research. *Rotunde Publikasjoner Rotunde, 84*, 55–76.

Slaughter, J. (2009). *Human rights, Inc.: The world novel, narrative form, and international law*. New York: Fordham University Press.

Slimak, M. W., & Dietz, T. (2006). Personal values, beliefs, and ecological risk perception. *Risk Analysis, 26(6)*, 1689–1705.

Slovic, P. (1987). Perception of risk. *Science, 236(4799)*, 280–285.

Slovic, P. (2000). *The perception of risk*. London: Earthscan Publishing.

Socolow, R. H., & Glaser, A. (2009). Balancing risks: Nuclear energy & climate change. *Daedalus, 138(4)*, 31–44.

Sokolowska, J., & Tyszka, T. (1995). Perception and acceptance of technological and environmental risks: Why are poor countries less concerned? *Risk Analysis, 15(6)*, 733–743.

Starr, S. (2016). Costs and consequences of the Fukushima Daiichi disaster. PSR's Environmental Health Policy Institute. Retrieved from https://www.psr.org/blog/resource/costs-and-consequences-of-the-fukushima-daiichi-disaster/

Stave, K. A. (2002). Using system dynamics to improve public participation in environmental decisions. *System Dynamics Review: The Journal of the System Dynamics Society, 18(2)*, 139–167.

Steiner, H. J., Alston, P., & Goodman, R. (2008). *International human rights in context: Law, politics, morals: Text and materials*. Oxford: Oxford University Press.

Steinhauser, G., Brandl, A., & Johnson, T. E. (2014). Comparison of the Chernobyl and Fukushima nuclear accidents: A review of the environmental impacts. *Science of the Total Environment, 470*, 800–817.

Steinhauser, G., Niisoe, T., Harada, K. H., Shozugawa, K., Schneider, S., Synal, H., … Koizumi, A. (2015). Post-accident sporadic releases of airborne radionuclides from the Fukushima Daiichi Nuclear Power Plant site. *Environmental Science & Technology, 49(24)*, 14028–14035.

Strupczewski, A. (2003). Accident risks in nuclear-power plants. *Applied Energy, 75(1–2)*, 79–86.

Sugimoto, T., Shinozaki, T., Naruse, T., & Miyamoto, Y. (2014). Who was concerned about radiation, food safety, and natural disasters after the great east Japan earthquake and Fukushima catastrophe? A nationwide cross-sectional survey in 2012. *PLoS One, 9(9)*, e106377. doi: 10.1371/journal.pone.0106377.

Sutou, Shizuyo. (2016). A message to Fukushima: Nothing to fear but fear itself. *Genes and Environment, 38(1)*, 12.

Suzuki, I., & Kaneko, Y. (2013). *Japan's disaster governance: How was the 3.11 crisis managed?* New York: Springer.

Suzuki, T. (2011). Deconstructing the zero-risk mindset: The lessons and future responsibilities for a post-Fukushima nuclear Japan. *Bulletin of the Atomic Scientists*, *67(5)*, 9–18.

Suzuki, T. (2015). Nuclear energy policy issues in Japan after the Fukushima nuclear accident. *Asian Perspective*, *39(4)*, 591.

Suzuki, Y., Takebayashi, Y., Yasumura, S., Murakami, M., Harigane, M., Yabe, H., ... & Maeda, M. (2018). Changes in risk perception of the health effects of radiation and mental health status: The Fukushima Health Management Survey. *International Journal of Environmental Research and Public Health*, *15(6)*, 1219.

Suzuki, Y., Yabe, H., Yasumura, S., Ohira, T., Niwa, S. I., Ohtsuru, A., ... & Abe, M. (2015). Psychological distress and the perception of radiation risks: The Fukushima health management survey. *Bulletin of the World Health Organization*, *93*, 598–605.

Syma Czapanskiy, K., & Manjoo, R. (2008). The right of public participation in the law-making process and the role of legislature in the promotion of this right. *Duke Journal of Comparative & International Law*, *19*, 1–40.

Taebi, B., & Roeser, S. (Eds.). (2015). *The ethics of nuclear energy*. Cambridge, MA: Cambridge University Press.

Tai, E. (2007). Korean ethnic education in Japanese public schools. *Asian Ethnicity*, *8(1)*, 5–23.

Takagi, E., & Saito, Y. (2015). Older parents' loneliness and family relationships in Japan. *Ageing International*, *40(4)*, 353–375.

Takahashi, S. (2014). *Radiation monitoring and dose estimation of the Fukushima nuclear accident*. Tokyo: Springer.

Takahashi, T. (2016). *Radiological issues for Fukushima's revitalized future*. Tokyo: Springer.

Tanaka, K., & Managi, S. (2016). Impact of a disaster on land price: Evidence from Fukushima nuclear power plant accident. *The Singapore Economic Review*, *61(1)*, 1640003.

Tanaka, N. (2013). Big Bang in Japan's energy policy. *Energy Strategy Reviews*, *1(4)*, 243–246.

Tateishi, Y. (2015). How should we discuss uncertainty in the context of environmental issues?: The health effects of radiation exposure from the Fukushima Daiichi Nuclear Disaster in Japan. *Japanese Sociological Review*, *66(3)*, 412–428.

Tateno, S., & Yokoyama, H. M. (2013). Public anxiety, trust, and the role of mediators in communicating risk of exposure to low dose radiation after the Fukushima Daiichi Nuclear Plant explosion. *The Journal of Science Communication*, *12(2)*, 1–22.

TEPCO. (2018). Compensation for nuclear damages. Retrieved from https://www7.tepco.co.jp/responsibility/revitalization/compensation-e.html

Thompson, T. J., & Beckerley, J. G. (1973). *The technology of nuclear reactor safety*. Cambridge, MA: MIT Press.

Thomson, R., Ito, N., Suda, H., Lin, F., Liu, Y., Hayasaka, R., ... & Wang, Z. (2012). Trusting tweets: The Fukushima disaster and information source credibility on

Twitter. In *Proceedings of the 9th international ISCRAM conference* (pp. 1–10). Vancouver: Simon Fraser University.

Thorme, M. (1990). Establishing environment as a human right. *Denver Journal of International Law and Policy, 19*, 301.

Toth, B. (2010). Public participation and democracy in practice-Aarhus convention principles as democratic institution building in the developing world. *Journal of Land, Resources, & Environmental Law, 30*, 295.

Tritter, J. Q., & McCallum, A. (2006). The snakes and ladders of user involvement: Moving beyond Arnstein. *Health Policy, 76(2)*, 156–168.

Tsubokura, M., Gilmour, S., Takahashi, K., Oikawa, T., & Kanazawa, Y. (2012). Internal radiation exposure after the Fukushima nuclear power plant disaster. *The Journal of the American Medical Association, 308(7)*, 669–670.

Turner, B. S. (1993). Outline of a theory of human rights. *Sociology, 27(3)*, 489–512.

UN. (1948). The universal declaration of human rights. Retrieved from http://www.un.org/en/universal-declaration-human-rights/index.html

UN. (1972). Declaration of the United Nations conference on the human environment. Retrieved from http://un-documents.net/unchedec.htm

UN. (1992). Rio declaration on environment and development. *Environmental Conservation, 19(4)*, 366.

Vadi, V. (2016). Energy security v. public health? Nuclear energy in international investment law and arbitration. *Georgetown Journal of International Law, 47(3)*, 1069–1197.

Valentine, S. V., & Sovacool, B. K. (2010). The socio-political economy of nuclear power development in Japan and South Korea. *Energy Policy, 38(12)*, 7971–7979.

Van der Pligt, J. (1992). *Nuclear energy and the public*. Malden: Blackwell Publishing.

Van der Pligt, J., Van der Linden, J., & Ester, P. (1982). Attitudes to nuclear energy: Beliefs, values. *Journal of Environmental Psychology, 2*, 221–231.

Veinla, I. H. (2005). Influence of the Aarhus convention on access to justice in environmental matters in Estonia. *European Energy and Environmental Law Review, 14*, 326.

Verestiuc, P. C., & Tucaliuc, O. M. (2015). Aspects concerning public access to environmental information. *Quaestus, 6*, 267–277.

Verschuuren, J. M., Etty, T. F. M., & Somsen, H. (2004). Public participation regarding the elaboration and approval of projects in the EU after the Aarhus convention. In *Yearbook of European environmental law*. Oxford: Oxford University Press.

Vick, D. W. (2002). The human rights act and the British constitution. *Texas International Law Journal, 37(2)*, 329.

Viljoen, F. (2007). *International human rights law in Africa*. Oxford: Oxford University Press.

Vivoda, V. (2012). Japan's energy security predicament post-Fukushima. *Energy Policy, 46*, 135–143.

Vivoda, V. (2016). *Energy security in Japan: Challenges after Fukushima*. London: Routledge.

Vivoda, V., & Graetz, G. (2015). Nuclear policy and regulation in Japan after Fukushima: Navigating the crisis. *Journal of Contemporary Asia*, *45(3)*, 490–509.

Walker, G. B. (2007). Public participation as participatory communication in environmental policy decision-making: From concepts to structured conversations. *Environmental Communication*, *1(1)*, 99–110.

Walker, J. S. (2004). *Three mile island: A nuclear crisis in historical perspective.* London, Berkeley, CA: University of California Press.

Wang, Q., & Chen, X. (2012). Regulatory transparency—How China can learn from Japan's nuclear regulatory failures? *Renewable and Sustainable Energy Reviews*, *16(6)*, 3574–3578.

Wanqing, C., Siwei, Z., & Hongmei, Z. (2014). Report of cancer incidence and mortality in China, 2010. *China Cancer*, *23(1)*, 1–10.

Watanabe, R. (2016). After the Fukushima disaster: Japan's nuclear policy change from 2011 to 2012. *Review of Policy Research*, *33(6)*, 623–645.

Watanobe, H., Furutani, T., Nihei, M., Sakuma, Y., Yanai, R., Takahashi, M., ... & Sagawa, F. (2014). The thyroid status of children and adolescents in Fukushima Prefecture examined during 20–30 months after the Fukushima nuclear power plant disaster: A cross-sectional, observational study. *PLoS One*, *9(12)*, e113804.

Wates, J. (2005). Aarhus convention: A driving force for environmental democracy. *Journal for European Environmental & Planning Law*, *2*, 2.

Weart, S. R. (2012). *The rise of nuclear fear.* Cambridge, MA: Harvard University Press.

Webster, T. (2009). International human rights law in Japan: The view at thirty, Columbia. *Journal of Asian Law*, *23(2)*, 244–266.

Whelan, D. J., & Donnelly, J. (2007). The west, economic and social rights, and the global human rights regime: Setting the record straight. *Human Rights Quarterly*, *29(4)*, 908–949.

White, P. (2015a). Japan's 2012 national debate on energy and environment policy. Retrieved from https://www.japanesestudies.org.uk/ejcjs/vol15/iss2/white.html

White, P. (2015b). Public participation in Japan's nuclear energy policy-forming process: Comparison of pre-and post-Fukushima processes (Doctoral dissertation). Retrieved from https://digital.library.adelaide.edu.au/dspace/handle/2440/98559

Whiteside, N., & Mah, A. (2012). Human rights and ethical reasoning: Capabilities, conventions and spheres of public action. *Sociology*, *46(5)*, 921–935.

Whitton, J., Charnley-Parry, I., Brasier, K., & Cotton, M. (2017). Shale gas governance in the United Kingdom and the United States: Opportunities for public participation and the implications for social justice. *Energy Research & Social Science*, *26*, 11–22.

Williams, D. (2002). Cancer after nuclear fallout: Lessons from the Chernobyl accident. *Nature Reviews Cancer*, *2(7)*, 543.

Wisadha, M. A. A., & Widyaningsih, G. A. (2018). Human rights and the environmental protection: The Naïveté in environmental culture. *Udayana Journal of Law and Culture*, *2(1)*, 73–96.

Wolfe, K. (2003). Greening the international human rights sphere-environmental rights and the draft declaration of principles on human rights and the environment. *Appeal: Review of Current Law and Law Reform, 9*, 45–58.

Woods, K. (2006). What does the language of human rights bring to campaigns for environmental justice? *Environmental Politics, 15(4)*, 572–591.

Woods, K. (2009). Suffering, sympathy, and (environmental) security: Reassessing Rorty's contribution to human rights theory. *Res Publica, 15(1)*, 53–66. doi: 10.1007/s11158-009-9083-5.

Woods, K. (2010). *Human rights and environmental sustainability*. Cheltenham: Edward Elgar.

Woods, K. (2014). *Human rights*. Basingstoke: Palgrave Macmillan.

Woods, K. (2016a). The rights of (future) humans qua humans. *Journal of Human Rights, 15(2)*, 291–306.

Woods, K. (2016b). Environmental human rights. In Teena, G., Hall, C., Meyer, J. M., & Schlosberg, D. (Eds.), *The Oxford handbook of environmental political theory* (pp. 333–345). Oxford: Oxford University Press.

World Nuclear Association. (2015). Radiation. Retrieved from http://www.world-nuclear.org/uploadedFiles/org/WNA/Publications/Nuclear_Information/pocket_guide_radiation.pdf

Xiao, Z., Shenghang, W., & Ehemann, E. M. (2017). Development of environmental rights in China: Substantive environmental rights or procedural environmental rights. *Frontiers of Law in China, 12(1)*, 24–56.

Xu, G., Feng, X., Li, Y., Chen, X., & Jia, J. (2017). Environmental risk perception and its influence on well-being. *Chinese Management Studies, 11(1)*, 35–50.

Yabe, H., Suzuki, Y., Mashiko, H., Nakayama, Y., Hisata, M., Niwa, S. I., ... & Abe, M. (2014). Psychological distress after the Great East Japan Earthquake and Fukushima Daiichi Nuclear Power Plant accident: Results of a mental health and lifestyle survey through the Fukushima Health Management Survey in FY2011 and FY2012. *Fukushima Journal Of Medical Science, 60(1)*, 57–67.

Yamakawa, M., & Yamamoto, D. (Eds.). (2016). *Unravelling the Fukushima disaster*. London: Routledge.

Yamamura, E. (2015). Transparency and views regarding nuclear energy before and after the Fukushima accident: Evidence on micro-data. *Pacific Economic Review, 20(5)*, 761–777.

Yamane, F., Ohgaki, H., & Asano, K. (2013). The immediate impact of the Fukushima Daiichi accident on local property values. *Risk Analysis, 33(11)*, 2023–2040.

Yamashita, S. (2014). Tenth Warren K. Sinclair keynote address—The Fukushima nuclear power plant accident and comprehensive health risk management. *Health Physics, 106(2)*, 166–180.

Yamauchi, T., Yoshikawa, T., Takamoto, M., Sasaki, T., Matsumoto, S., Kayashima, K., & Takahashi, M. (2017). Overwork-related disorders in Japan: Recent trends and development of a national policy to promote preventive measures. *Industrial Health, 55(3)*, 293–302.

Yasunari, T. J., Stohl, A., Hayano, R. S., Burkhart, J. F., Eckhardt, S., & Yasunari, T. (2011). Cesium-137 deposition and contamination of Japanese soils due to the Fukushima nuclear accident. *Proceedings of the National Academy of Sciences of the United States of America, 108(49)*, 19530–19534.

Yildiz, K., Baykal, T., & Altin, M. (2002). Çevrenin tanınması ve öneminin kavranmasına yönelik örnek bir sulak alan çalışması. *Gazi Üniversitesi Gazi Eğitim Fakültesi Dergisi, 22(3)*, 1–9.

Yilmaz, Senol. (2011). *Fukushima nuclear disaster: A study in poor crisis communication. (RSIS Commentaries, No. 093). RSIS Commentaries*. Singapore: Nanyang Technological University.

Yontrarak, Pat. (2018). The Fukushima Daiichi accident: Water contamination in Japan. Retrieved from http://large.stanford.edu/courses/2018/ph241/yontrarak2/

Yoshida, K., Shinkawa, T., Urata, H., Nakashima, K., Orita, M., Yasui, K., … & Takamura, N. (2016). Psychological distress of residents in Kawauchi village, Fukushima prefecture after the accident at Fukushima Daiichi nuclear power station: The Fukushima health management survey. *Peerj, 4*.

Yoshida, N., & Kanda, J. (2012). Tracking the Fukushima radionuclides. *Science, 336(6085)*, 1115–1116.

Yoshida, N., & Takahashi, Y. (2012). Land-surface contamination by radionuclides from the Fukushima Daiichi Nuclear Power Plant accident. *Elements, 8(3)*, 201–206.

Zarsky, L. (2012). *Human rights and the environment: Conflicts and norms in a globalizing world*. London: Earthscan.

Zhang, Q., & Mclellan, B. C. (2014). Review of Japan's power generation scenarios in light of the Fukushima nuclear accident. *International Journal of Energy Research, 38(5)*, 539–550.

Ziemer, L. S. (2001). Application in Tibet of the principles on human rights and the environment. *Harvard Law School Human Rights Journal, 14*, 233.

Index

About the Author

Emrah Akyüz is a lecturer in the Department of Management and Organization at the University of Bingol in Turkey. He holds an MSc in International Development from the University of Edinburgh in 2014 and a PhD in Environmental Law from the University of Leeds in 2019, where his doctoral and master's research theses were funded by the Republic of Turkey. He is an environmental social scientist conducting research in the area of energy, human rights, and environmental policies, environmental human rights, environmental law, and urban politics. His research is regionally focused in Turkey and Japan. He has published more than fifteen journal articles related to his research interests. These papers have been subsequently reported by hundreds of newspapers, news channels, magazines, and news websites in Turkey. He is also the author of three books which include: (1) *100 Unknown Scientific Facts in Environmental Problems*, (2) *Save the World: The Scientific Secrets of the Clean Environment*, (3) *The ABC of Environmental Science: Environmental Issues and Policies for Beginners*.

www.ingramcontent.com/pod-product-compliance
Ingram Content Group UK Ltd.
Pitfield, Milton Keynes, MK11 3LW, UK
UKHW041843031026
15587UKWH00002B/971

9 781793 637819